# ISLAND BORN OF FIRE

# ISLAND BORN OF FIRE:

## Volcano
## Piton de la Fournaise

**Written & Illustrated By:**

**R. B. Trombley, PhD**
Southwest Volcano Research Centre
Apache Junction, Arizona

iUniverse, Inc.
New York  Lincoln  Shanghai

## ISLAND BORN OF FIRE:
### Volcano Piton de la Fournaise

iUniverse books may be ordered through booksellers or by contacting:

iUniverse
2021 Pine Lake Road, Suite 100
Lincoln, NE 68512
www.iuniverse.com
1-800-Authors (1-800-288-4677)

Originally Published by

McGraw-Hill Publishing Co.
March, 1996

ISBN-13: 978-0-595-41775-9 (pbk)
ISBN-13: 978-0-595-86118-7 (ebk)
ISBN-10: 0-595-41775-2 (pbk)
ISBN-10: 0-595-86118-0 (ebk)

Printed in the United States of America

Respectfully Dedicated to Mom & Dad, Sister Karen, the people of Réunion Island and all geoscientists throughout the world.

# TABLE OF CONTENTS

# APPENDIX A
# PHOTOGRAPHIC PLATES

# LIST OF ILLUSTRATIONS

# LIST OF TABLES

# PREFACE

Certainly one of the most spectacular events in nature is an erupting volcano. It is spectacular. Volcanoes can be beautiful, informative, and both beneficial and dangerous. They serve as windows through which we can dimly perceive the interior of the earth. Volcanism occurs in many different styles, principally near tectonic plate boundaries and also, as is the case with Piton de la Fournaise, in the middle of an ocean. Piton de la Fournaise is one of the most active volcanoes in the world today. There is something to be said about the beauty of volcanoes and Piton de la Fournaise is no exception. Piton de la Fournaise contributes to the well being of its tropical island home, Réunion Island. The atmosphere and the oceans originated in volcanic episodes of the distant past and Piton de la Fournaise was born of fire and from the Indian Ocean. Soils derived from the volcanic materials of Piton de la Fournaise are exceptionally fertile on the island. Emissions of volcanic rock, gas, and steam are also important sources of industrial materials and chemicals, such as pumice, boric acid, ammonia, and carbon dioxide. All these elements serve and help support the island.

It is my hope that the following work will serve as an insight to this beautiful volcano and this beautiful island. Yes, consider the volcano Piton de la Fournaise. It is nature. It is powerful. It is awesome. It is fire. It is spectacle. It is beauty. It is life. It can be (but has not yet been) death.

R. B. Trombley
January, 1996

# CHAPTER 1

# RÉUNION ISLAND

## Historical And Cultural Aspects Of Réunion Island

It is a beautiful volcanic island. It is full of colour, splendor, nature and life. (See Plates in Appendix A.) The people are warm, friendly and, of course, very French. Réunion Island, located in the Indian Ocean, is currently believed to have been discovered early in the 16th century by the Portuguese navigator Pedro de Mascarenhas. In the year 1638, Réunion Island was claimed by France as a stopover point for mercantile ships on their way to India. A small colony was established on Réunion in the year 1665 by the French East India Company. Originally called Île de Bourbon, it was named Réunion in 1793 during the French Revolution.

Great Britain held the island from the years 1810 to 1814, when it was then returned to France. It was renamed Bourbon until 1848, when the name was again changed back to Île de la Réunion. The colony was made an overseas department of France effective 1 January 1947.

The island of Réunion, located in the Indian Ocean, constituting an overseas department of France, and is the most westerly of the Mascarene Islands that also includes the islands of Mauritius and Rodriguez. Réunion Island is situated approximately 690 kilometres (430 miles) east of Madagascar. Réunion Island is approximately 72 kilometres long and 52 kilometres wide (45 by 32 miles) with an area of 2,512 square kilometres (970 square miles). The capital of Réunion Island is the town of St. Denis with a population just over 103,500. Other principal towns on Réunion Island are St. Paul, St. Pierre, St. Louis, and St. Joseph.

As an overseas department of France, Réunion Island sends three deputies and two senators for the French National Assembly. The local government of Réunion Island is a prefect, an elected general council, consisting of 36 members, and an elected regional council consisting of 46 members.

Réunion Island is totally of volcanic origin and consists largely of rugged mountains and high plateaus enclosed by a narrow coastal plain. The highest peak on the island is the extinct volcano, Piton des Neiges (Snowy Peak) at 3,070 metres (10,070 feet). The current active volcano is Piton de la Fournaise (Furnace Peak) at 2,625 metres (8,612 feet) and is located in the southeast volcanic region of the island.

Réunion Island's climate ranges from hot and humid in the coastal lowlands to temperate in the higher elevations. The hottest months, November through April, are also the wettest months. Rainfall, accompanied by the southeasterly trade winds is heaviest on the eastern (windward) coast, whereas the southwestern side of the island remains dry.

The economy of Réunion Island is agricultural in nature, primarily based on sugar, which constitutes more than 80% of its exports. Other products produced from the sugarcane include rum and molasses. Réunion Island also exports other products to help support the economy. Those exports include tobacco, vanilla, manioc, tapioca and essential oils of geranium and vetiver. Réunion Island also imports some products. These imports consist primarily of foodstuffs, cement, iron and steel products. The principal port for the import/export trade is Le Port, at Pointe des Galets located on the northwest coast. Almost all of Réunion Island's foreign trade is with France and Madagascar.

There used to be a coastal railroad that links Le Port with St. Denis at the northern tip of the island and with St. Benoît and St. Pierre but it has long since been discontinued in favour of the automobile. Airlines as well as ships connect Réunion Island Madagascar, the African mainland and Europe. The well-maintained highway network throughout Réunion Island now links the principal cities.

Réunion Island's population, estimated at 550,000 is largely Creole, descendants of the original French settlers. The official language of Réunion Island is French. Although French is the official language, there has been considerable intermingling with the Bantu speaking Malaguary and Kaffir peoples of southern Africa, as well as with immigrants from India, Indochina, China, and East Africa, many of who arrived as indentured labour for the many plantations of sugar and coffee.

It is interesting to note that there are some interesting side benefits for the volcanologists, guest volcanologists, geophysicists, scientists and the like who are visiting Réunion Island. The people of Réunion Island hold a very high regard for all those scientists associated with L'Obervatoire Volcanologique du Piton de la Fournaise and/or L'Institut de Physique du Globe de Paris (IPGP) whose headquarters are in Paris, France. The people of the villages know that the scientists of Réunion Island are constantly monitoring *their* mountain, *their* volcano. This is done 24 hours a day, 365 days a year. The villagers know the risks taken by the scientists. The villagers are warned of impending eruptions in time to scurry to safety if warranted. Not only the safety of themselves, but the safety of their animals and, in some cases, their properties. To date, not a single life has been lost to an eruption of Piton de la Fournaise. So great is the admiration for the volcanologists, geophysicists, and scientists of Piton de la Fournaise, that when any of the scientists appears at a restaurant in any one of the many villages about the island, that, if there was "…no seating available at the moment", there magically is immediate seating available for those associated with L'Obervatoire Volcanologique du Piton de la Fournaise and/or L'Institut de Physique du Globe de Paris (IPGP).

A recent compilation of the submarine data around Réunion allows, for the first time ever, to describe the entire volcanic system of Réunion. The data for such an undertaking originated from the general bathymetric map of Réunion and from detailed surveys from cruises on the east flank of the island. The general shape of Réunion Island is that of a flattened cone with a base diameter of 220 kilometres (136.7 miles) and a height of 7 kilometres (4.35 miles).

The emerged portion of the island represents only 1/32nd of the total volume of Réunion Island and a mere 1/100th if the material buried in the lithosphere flexure is taken into account. Three main zones are distinguished: the volcanic emerged and submarine zones, and the lower slopes that are now covered by sedimentary layers. The subaerial-subaquaeous transition is marked by a sharp downward-slope increase corresponding to the change of flow conditions between the two mediums. This feature of the island is observed near the seashore in the youngest zones, and is submerged in the oldest zones as a result of the subsidence of the island since the termination of activity, principally of Piton des Neiges, in those areas.

Detailed studies eastward of Piton de la Fournaise indicate the presence of a volume of about 500 cubic kilometres (1640 cubic feet) of mass wasting deposits. Virtually all of the material consists of sub-aerially erupted, fragmented lavas with ages ranging between 10 and 100 thousand years. The interpretation of bathymetry and acoustic imaging is highly suggestive of several debris avalanches and slumps.

The obvious morphologic similarity between virtually all of the flanks of Réunion Island and the east flank, suggest that mass wasting deposits are present almost everywhere and are considered commonplace.

## Morphological Features Of The Réunion Island

The Réunion Island is made up of two great mainly basaltic and juxtaposed volcanoes: in the northwest, the extinct Piton des Neiges, 3069 metres (10,069 feet) and in the southeast, the still active Piton de la Fournaise, 2631 metres (8,632 feet). Located in the south tropical zone, it is subjected to a southeast trade wind regime which earns it a very wet climate. Its morphoclimatical and structural conditions were at the origin of specific reliefs which are common in other similar volcanic areas such as the Hawaiian Islands.

Piton des Neiges was initially a submarine volcano before its emersion, more than 2.1 million years ago. Its activity ended about twenty thousand years ago. It is distinguished by noteworthy planes which constitute its external slopes. Its central area is smashed down by three great cirques set as cloverleafs around the complex built by the last phases of activity. The cirques were opened by erosion, which has worked from a huge summit caldera ("Caldera II") formed perhaps 150,000 years ago, after very explosive episodes. Their digging is later than the last eruptions. It was accompanied by the exportation of big quantities of alluvial material the deposit of which, next to the sea, has built several wide alluvial cones.

Piton de la Fournaise was built on the southeastern slopes of Piton des Neiges, since more than half a million years. Its slopes are already cut by deep valleys. It summit areas show a noteworthy fitting of calderas, whose last is the Enclos in the middle of which stands the present terminal cone. Its eastern slopes were affected by landslides, the more recent, the Grand Brulé (Big Burn), constitutes, with the Enclos, a morphostructural "U-shape" opened towards the Indian Ocean. The terminal cone is topped by a one kilometre wide crater complex (Bory and Dolomieu Craters), in continual evolution. The last eruptions have produced the whole range of shapes which characterise "Hawaiian" volcanism.

Present landscapes result from interaction between volcanism and erosion. The strength of the erosion has been considerable. It has appeared since ancient periods. Its recent action seems most spectacular because it has dug the cirques of Piton des Neiges and some valleys of Piton de la Fournaise in less than about twenty thousand years.

Detailed studies eastward of Piton de la Fournaise show the presence of a volume of about 500 cubic kilometres of mass wasting deposits. Virtually all of the material consists of subaerially erupted, fragmented lavas ranging in age between about 10 and 10,000 years (sampling and dating are not exhaustive). The interpretation of the bathymetry and of the acoustic images strongly suggest the existence of several debris avalanches and slumps.

The obvious morphologic similarity between virtually all flanks of the island and the east flank suggests that mass wasting deposits are present almost everywhere. These observations could lead to a re-examination of the models of evolution of Piton des Neiges and Piton de la Fournaise.

## Submarine Morphologies And Structures Of Réunion

A compilation of the submarine data around Réunion Island allows, for the first time, to describe the whole volcanic system of Réunion Island The data has originated from the general bathymetric map of the island (See Figure 1–1), and from the detailed surveys of Fournaise 1 and 2 cruises, on the east flank of the island. The general shape of Réunion Island is that of a flattened cone with a base diameter of 220 kilometres (136.7 miles) and a height of 7 kilometres (4.35 miles). As has been previously mentioned, the emerged portion represents only 1/32 of the volume of the island, and merely 1/100 if the material buried in the lithospheric flexure is taken into account. Three main zones are distinguished: the volcanic emerged and submarine zones, and the lower slopes which are covered by layered sediments. The subaerial-subaqueous transition is marked by a sharp downward-slope increase corresponding to the change of flow conditions between the two mediums. This feature is observed near the seashore in the youngest zones, and is submerged in the oldest zones as a result of the subsidence of the island since the termination of activity in those areas.

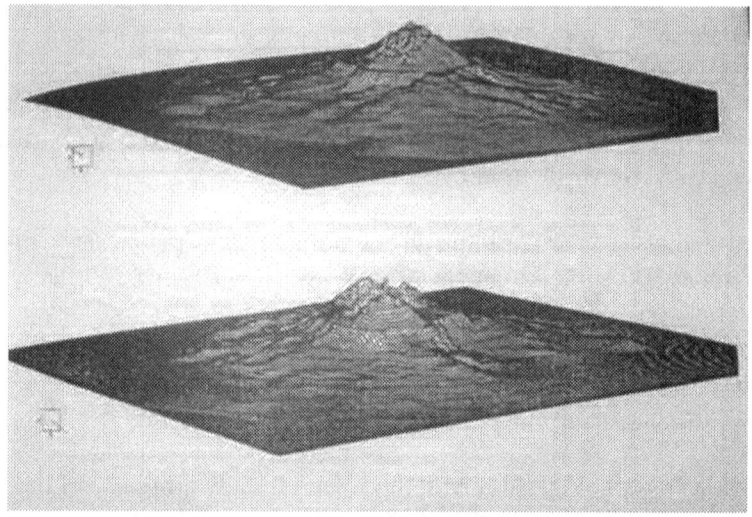

*Figure 1–1. General bathymetry and topography (100 m contour)*

# CHAPTER 2

# HOT SPOT VOLCANOES

Students of the science of volcanology understand that there is a correlation between the edges of the moving tectonic plates of the Earth's crust and volcanoes. But what about the volcanoes that occur within the tectonic plates themselves. The famous Hawaiian Islands, where some of the world's most active volcanoes are located, are right in the middle of the Pacific plate. Approximately 95 percent of all the world's active volcanoes lie along the edges of tectonic plates. The other 5 percent are located within the plates themselves. But why? What is the mechanism that causes this 5 percent of the world's volcanoes to be located here? Piton de la Fournaise is one of those volcanoes that lies within a tectonic plate; namely, the Africa plate.

There are some fundamental differences between volcanoes that lie along the edges of tectonic plates and those volcanoes that reside within a tectonic plate. One of those differences is the topography or physical profiles of the terrain. Figure 2–1 illustrates those differences. Volcanic islands such as the Seychelles Mauritius Ridge that has the islands of Kerguelan, Coetivy, Tromelin and Réunion, which has the currently active volcano, Piton de la Fournaise, all lie within a plate. The Hawaiian Islands are another example of the formation of a chain or string of islands that are spread out over a large distance to form such island chains or arcs.

Although most volcanic activity is concentrated at plate boundaries, volcanoes within plates can produce enormous volcanoes of lava. The example stated previously is the active volcano chain that begins with Hawaii and continues as a string of progressively older, extinct, eroded, and submerged volcanic ridges and mountains. Except for Hawaii itself, where earthquakes are triggered by volcanic activity, the chain is essentially aseismic (without earthquakes) and is called an aseismic ridge different from the seismic mid-ocean ridges where seafloor spreading takes place.

Aseismic ridges, which also occur elsewhere in the Pacific and in the other large oceans, were difficult to incorporate into the framework of plate-tectonic theory. Jason Morgan of Princeton and J. Tuzo Wilson of Toronto introduced the concept of "hot spots" to explain aseismic ridges and other volcano centres within continents. They propose that "hot spots" are the surface manifestations of jets or plumes of hot material that rise from deep within the mantle, drill through the lithosphere, and emerge as volcanic centres (See Figure 2–2). These columnar currents are supposedly fixed in the mantle and do not move with the plates. As a result, the "hot spot" leaves a trail of extinct, progressively older volcanoes as the plate moves over it. If "hot spots" are indeed fixed in the mantle, the trail of volcanoes carried away from the "hot spot" provides a powerful method of measuring the velocity and direction of plate motion. The assumption that "hot spots" are immobile is now being examined very carefully by geophysicists. In a general way the "hot spot" origin of the Hawaiian and Emperor Seamount Chains has been confirmed by deep-sea drilling which has established the progressively older periods of volcanism along the chains.

A second type of ocean-bottom formation, extinct undersea basaltic volcanoes rising kilometres or more above the sea floor, are common intraplate volcanic features. More than 10,000 are known in the Pacific Ocean alone. They are thought to have originated as active volcanoes near spreading centres and to have become extinct as the plate carried them away. Some initially projected above sea level but have been eroded to flat-topped cones by wave action. As the oceanic lithosphere spread, cooled, and contracted, the sea floor subsided, dropping the truncated volcanoes below sea level. These are known as guyots.

Volcanic centres within the continental plates take different forms. Fissure eruptions of basalt, like those in the Pacific Northwest of the United States, suggest that a fracture penetrated the continental lithosphere and that basalt spurted rapidly to the surface without much contamination from the felsic crust. Perhaps fractures developed in the early stages of an abortive spreading episode. Eruptions of basalt that mark the initial stages of continental rifting, or breakup, can be documented in several parts of the world. For example, basalt flows are found today in troughs in eastern North America bounded by ancient faults. These flows formed at the beginning of the breakup of Pangaea some 200 million years ago. Basalt is also found in association with the rift valleys of East Africa—a feature that some geologists interpret as representing an abortive stage in the breakup of that continent.

The vast ignimbrite sheets, which are the rhyolitic counterpart of flood basalt, have no easy explanation. Perhaps magma rising slowly from the upper mantle assimilates large quantities of the siliceous granitic or sedimentary crust and in this way is transformed into rhyolitic magma.

Another fundamental difference between volcanoes that lie along the edges of tectonic plates and those volcanoes that reside within a tectonic plate is the age distribution of the two. It is a well known fact in geology that there is a definite age differential in volcanic island arcs. The Hawaiian arc, for example, stretches from the big island of Hawaii to the island of Midway some 2500 km (1553 miles) northwest. Radioactivity dating has shown that the age of the volcanic rocks becomes progressively older as measured from the big island of Hawaii towards the island of Midway. The islands of the Seychelles Mauritius Ridge chain and surrounding areas also show an age differential from one end to the other.

The age distribution patterns of volcanoes of the Seychelles Mauritius Ridge is very much different from that of tectonic plate margin volcanoes. In tectonic plate margin volcanoes, as well as mid-oceanic ridge volcanoes, young volcanoes form along the edges of the plates near the subduction zones. In mid-oceanic ridges, indications are that there is progressive aging on either side of the ridge as one traverses outward from the ridge.

As previously stated, associated with volcanoes is earthquake phenomena. Even the pattern of earthquakes is much different and exhibits a different pattern along subduction zones as opposed to inter-plate volcanic zones. Earthquakes occur along the entire length of subduction zones whereas earthquakes appear to concentrate under or near the active volcano within a tectonic plate zone.

Plotting the location of earthquakes at depth under a tectonic plate located at a volcano, outlines the conduits and fissures through which magma forces its way to the surface. Buoyancy is the providing force of the lighter magma which is surrounded by denser, crystalline rock. As the magma rises, the more dense surrounding rocks are cracked open, creating fissures, by changing pressure and temperatures. This is the cause of the earthquakes.

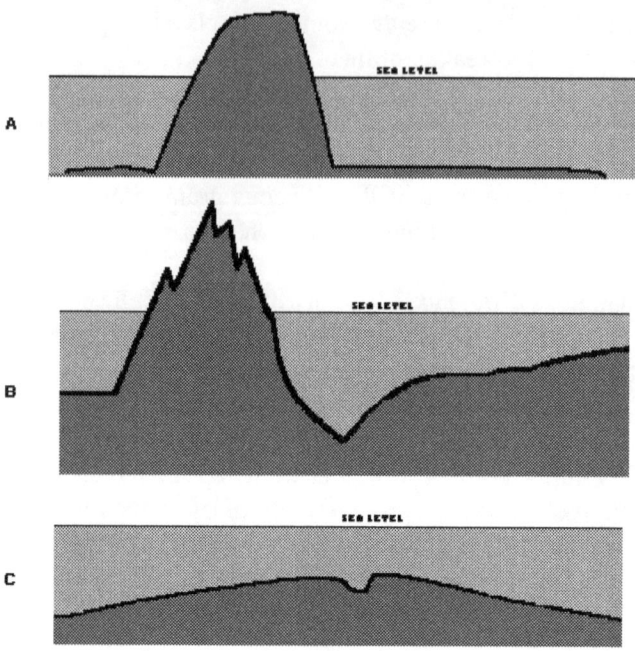

*Figure 2–1.   Comparative topographic profiles of (A) an island chain
or arc ridge, (B) a subduction zones, and (C) mid-ocean
ridges.*

As previously mentioned, A Canadian geophysicist, J. Tuzo Wilson, developed the "hot-spot" idea as an explanation for the anomalous position of mid-
plate volcanoes and yet relates them to plate tectonics. This type of volcanism
is linked to upwellings of hot materials coming from deep within the mantle,
which upon reaching the base of the lithosphere, pierces it like a hot needle
through butter. The surface expression of the column or plume is called a "hot-
spot". This phenomena gives rise to a type of volcanism whose initial magma
is an alkaline, relatively fluid basalt that erupts at the Earth's surface as a lava
flow. Figure 2–2 illustrates this phenomena.

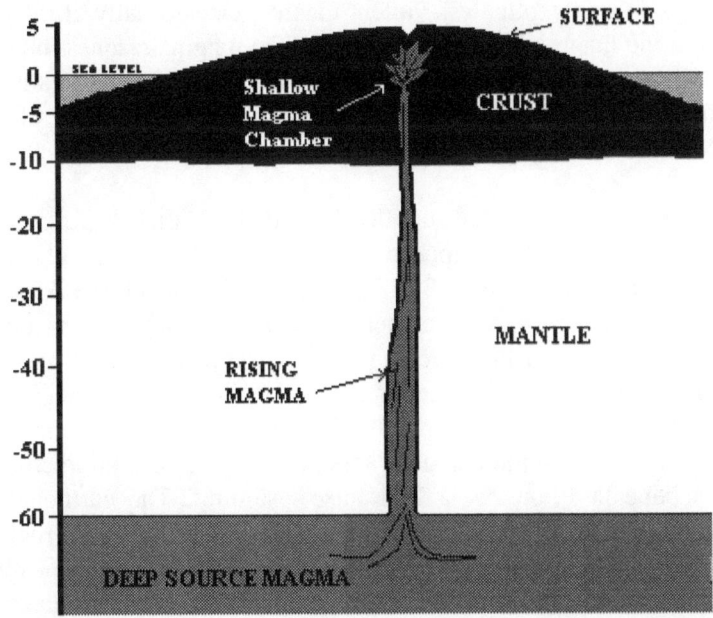

*Figure 2–2. Profile of a "hot-spot" volcano and its source of magma.*

This column or plume is considered stationary in relation to the core of the Earth, and the displacement of a tectonic plate above this plume leaves a trace on the crust in the form of strings of volcanic arcs or chains. Currently, there are about a hundred known "hot-spots" about the globe.

"Hot-spot" volcanoes are of great interest and importance to volcanologists because their frequent eruptions make for fantastic natural observatories and make excellent sites for refining observational and forecasting techniques. As new discoveries are made, it becomes increasingly more evident that the studies of volcanoes and plate tectonics are closely interwoven. The plate tectonic concept explains the location of volcanic belts while the tracks of extinct volcanoes lend clues to the motion and direction of ancient tectonic plate motions.

Almost all "hot-spot" volcanoes are effusive, that is, they erupt streams of molten lava flows. The flows generally consist of solid, hot fragments. The characteristics of effusive eruptions are that the gas content in the magma is low and the magma itself is of relatively low viscosity. This allows the gases to boil out in a more gentle, less violent manner. Occasionally, if the trapped gases within the magma have escaped before the magma reaches the surface, the lava will quietly well out of the vent as an incandescent stream molten rock. Effusive eruptions have a Volcanic Explosive Index (VEI) of 0 to 2. Appendix B describes the VEI in detail.

The deep source of magma that supplies Piton de la Fournaise and other "hot-spot" volcanoes of this type, appear to be within the so-called plastic layer which lies 50 to 60 kilometres (31.1 to 37.3 miles) beneath the Africa plate and is also the maximum depth of detectable earthquakes. The shape of the earthquake zone beneath Piton de la Fournaise indicates that of an inverted cone, approximately 1 to 2 kilometres (.6 to 1.2 miles) in diameter and 50 to 60 kilometres (31.1 to 37.3 miles) high. This allows the deep magma chamber to supply a shallow magma chamber approximately .5 to 2 kilometres (.3 to 1.2 miles) beneath Piton de la Fournaise's summit. The earthquakes are thought to have their origins in the slipping and faulting of brittle rock material that suddenly fails by fracturing. The plastic or molten rock within the chambers can easily deform and therefore produces no earthquakes. The magma chamber model is supported as evidenced by the tilt patterns exhibited around Piton de la Fournaise and other "hot-spot" volcanoes like it. This push of magma inflates the volcano prior to erupting.

A model of both mechanical and thermal behaviour of the lithosphere within a zone across the Indian Ocean and centred on Réunion Island has been worked out. It is based on both bathymetric and geoid data. As it turns out, the zone belongs entirely to the African plate and is bounded by Madagascar on the west and by the Central and Southwest Indian ridges on the east and south respectively (See Figure 2–3). It comprises the Madagascar basins, and the Southern Marcarene Plateau composed of the Nazareth and Cargados-Carajos banks and by the volcanic island of Mauritius According to a recent article by Jean-Christophe Sempéré and Emily M. Klein, in *EOS*, the transactions of the *American Geophysical Union*, the Southwest Indian Ridge (SWIR) separates the Antarctic and African plates and extends from the Bouvet Triple Junction in the South Atlantic to the Rodriguez Triple junction in the Central Indian Ocean (See Figure 2–4). One of the most important characteristics of the SWIR, from a global perspective and relative to the Réunion Island complex, is that it is associated with extremely slow spreading rates 13–18 mm/yr (.042-.059 ft/yr). The two triple junctions bounding the SWIR differ in their characteristics: the relatively stationary Bouvet Triple Junction is associated with a "hot spot", while the Rodriguez Triple Junction is actively migrating. The SWIR axis is marked by a rift valley, the dimensions of which vary along-strike in a poorly understood manner. It also displays variations in segmentation geometry along-axis with long sections that are very linear and orthogonal to the spreading direction and sections that are segmented by transforms at intervals of 50–100 kilometres (31.1 to 62.1 miles) and oblique to the spreading direction. The SWIR is influenced by its proximity to the Bouvet and Marion "hot spots". In 1996, a field programme will explore an unmapped section of the SWIR between 15°E and 35°E. The objectives of that geophysical study are to characterise crustal accretion and segmentation in two contrasting domains one consisting of linear ridge segments and one dominated by transform faults. Several important scientific problems can be addressed by studying the SWIR. Most geophysical and geochemical studies to date have focused on the slow spreading Mid-Atlantic Ridge, intermediate spreading ridges such as the Juan de Fuca Ridge, or the fast spreading East Pacific Rise. Little is known about the crustal or lithospheric structure, the chemical systematics of magmatism, or the nature of hydrothermal activity at the slowest end of the spreading spectrum such as that found in the SWIR.

Theoretical work has suggested that the style of crustal accretion and magmatism may differ significantly at very slow spreading rates, although there are few data to test these models. From a geochemical standpoint, because crust forming along some portions of the SWIR may be produced by very small degrees of melting, the rocks recovered may preserve a clearer record of the variability in upper mantle source composition. In addition, because the crust along portions of the SWIR is likely to be thinner than normal, the large topographic relief of the rift valley and transform faults along the SWIR should offer an excellent opportunity to sample the lower crust and upper mantle and compare their compositions to spatially related basalts. Contrasting styles of lithospheric accretion and deformation at triple junctions can also be studied at the SWIR.

The topography of the ridges such as the Mascarene Ridge is controlled by the loading and bending effects of the volcanic chain on the oceanic lithosphere and by the rejuvenation of the plate due to underplate re-heating (Crough, 1978). Since the elastic deformation of the lithosphere depends on its thermal properties at the time of loading (Watts, et al, 1980), the loading effects of three main zones (i.e., the Southern Mascarene Plateau, the Mauritius and Réunion islands) have been considered. The highest rigidity occurs under

Réunion Island and thereby suggests a recent "hot-spot" origin for the volcanism 3 to 5 million years ago. The evolution of the rigidity and thus the age of
the lithosphere at loading time along the ridges is within the concept frame of
the "hot-spot" theory. The regional upwelling in the Mascarene basin is
explained by the conductive heating of a plate moving over a deep thermal
source using the model proposed by Nakiboglu and Lambeck (1985), this
mantle source having probably initiated the initial volcanism at Réunion
Island.

Figure 2–5 illustrates a trail of lavas across the Indian Ocean and traces the
location of the "hot-Spot", which stays put as the tectonic plates moves over
it. The speed and direction of travel of the ocean floor can be discovered from
the sequence of ages of the lavas erupted onto it from the "hot-spot". The
island of Réunion lies above the "hot-spot" at this and the volcano Piton de la
Fournaise, located on Réunion Island, is the currently active volcano as volcanoes Mona Loa and Kilauea are currently the active volcanoes of the
Hawaiian Island chain.

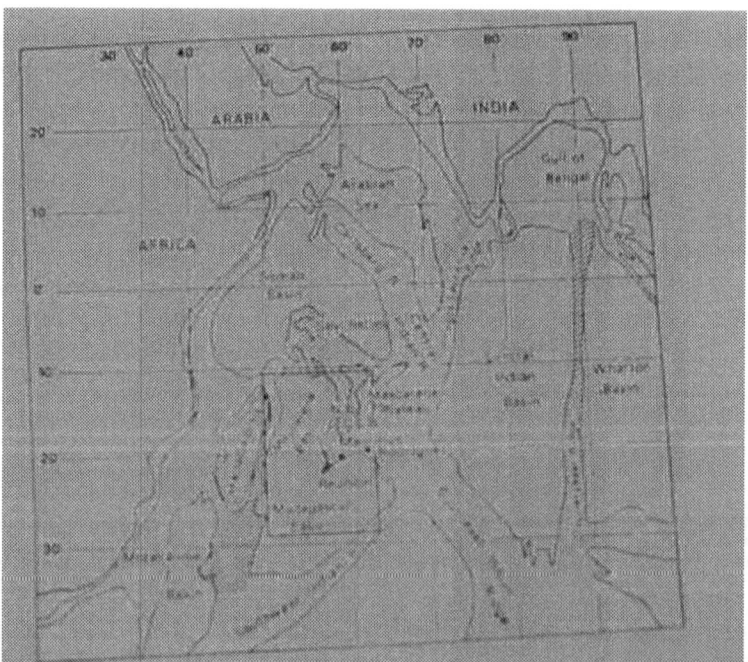

*Figure 2–3. Structural Sketch Map Of The Indian Ocean*

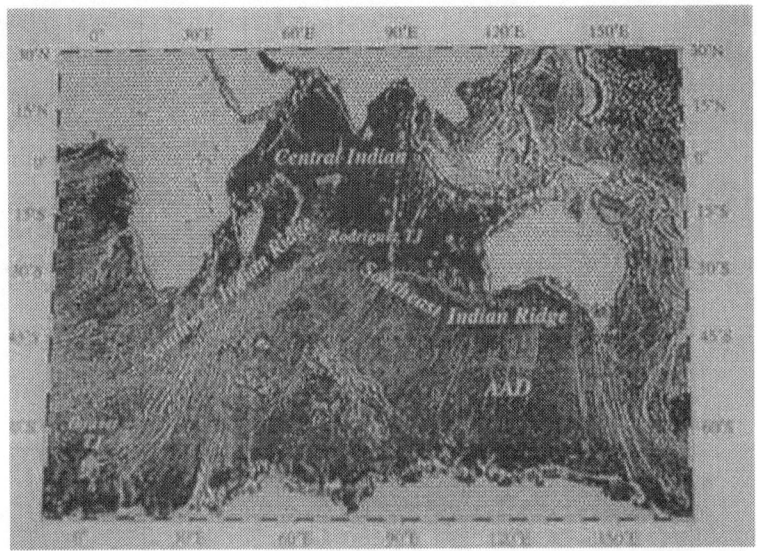

*Figure 2–4.   Satellite Derived Gravity Field Over The Indian*
*Ocean Showing The Three Major Spreading Centres*
*Including The Southwest Indian Ridge (SWIR)*

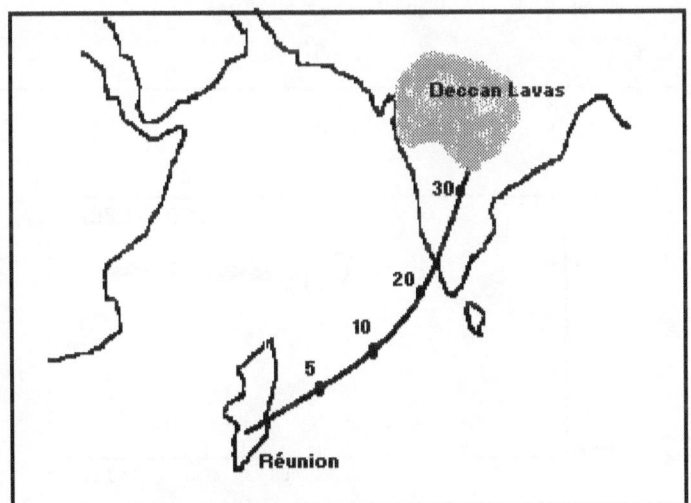

*Figure 2–5. Movement of a tectonic plate over a "hot-spot".*

Table 2–1 along with Figure 2–6 illustrates further the stationary "hot-spot" model showing that the heat source is emplaced at the base of the oceanic lithosphere. The velocity of the plate, relative to a fixed mantle "hot-spot", is taken to be 2.5 centimetres/year (approx. 1 inch/year).

This particular "hot-spot" is still one of the most prolific active volcanic sites on Earth. Over the last two million years, Réunion Island has built into a large volcanic island, whose summit stands at 7 kilometres (22,966 feet) above the sea-bed.

*Table 2–1. Absolute Speed Of The African Plate*

| Period millions of years ago | Absolute Movement of the African Plate (Duncan, 1981) | | Absolute Speed Over The Reunion "Hot-Spot" cm/yr |
| | Poles of Rotation Lat.          Long. Angle | | |
| 0 -21 | $61.0^0$ N          $45.0^0$ W $4.6^0$ | | 2.3 |
| 0 - 36 | $42.5^0$N          $25.0^0$W $9.5^0$ | | 2.9 |

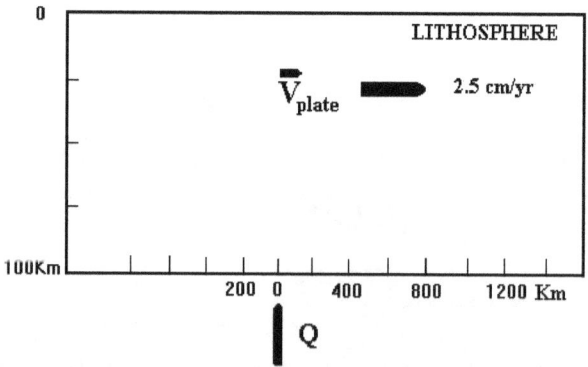

*Figure 2–6. Stationary "hot-spot" model.*

# CHAPTER 3

# VOLCANO PITON de la FOURNAISE

## Introduction

Piton de la Fournaise, 2635 metres (8644 ft.), is the active shield volcano of Réunion Island ($55^\circ$ 43'E, $21^\circ$17'S) in the western Indian Ocean. It is one of the most regularly active volcanoes in the world, averaging one event every 10 months. Its mean effusion rate has been calculated at 0.3 m/s (.98 ft/s) during the last 60 years. It last erupted in August of 1992.

The oldest data on Piton de la Fournaise's historical activity, given by sailors, date from the year 1640. A complete history of Piton de la Fournaise's eruption history may be found in Appendix C. Archives and other investigations have brought to light about 151 subaerial events. No historical or prehistoric submarine eruptive centres are known.

## Volcano-Structural Evolution Of Piton de la Fournaise During The Last .53 Million Years

Owing to the recent field studies in the great valleys of Piton de la Fournaise volcano and considering new geochronologic data, it can be determined that there has been a succession of volcano-structural events in the growing of the volcano Piton de la Fournaise since .53 million years ago. The oldest series are principally described herein.

Two major tectonic events seems to have marked the edification of volcano Piton de la Fournaise during this period. The first, dated approximately

300,000 years ago, corresponds to the stoppage of volcanic activity in the south-southwest part of the massif and to the existence of the first collapse structure, the Rivière des Remparts caldera, for which the western boundary corresponds approximately to the upper part of the Rivière des Remparts. The pre-collapse formations are named série ancienne (ancient series). Two periods are petrographically distinguishable, the période des "laves pintades" (period of the pintade lavas) and the période des "laves à olivines" (period of the olivine lavas). The boundary between these two periods remains indeterminate.

The second event, dated approximately 150,000 years ago, appears to be char-acterised by the displacement of the central zone of the volcano from the Plaine des Sables area to its present position. A new collapse event, the Morne Langevin caldera, is associated with this displacement; the post-Morne Langevin lavas and are named "bouclier récent" (recent shield). The intermedi-ate period between these two major events named "série du volcan de Mahavel" (Mahavel volcanic series) is marked, on the western flank of the vol-cano, by the opening of an ancient valley named paleo-Rivière des Remparts. The existence of this erosive structure is attested by the presence of many mud-flows (lahars) in the walls of the present valley. During this time, the volcanic activity continued at a low rate in this sector, corresponding to the functioning of a volcano centre that remained at the same place during the série ancienne. So, the série ancienne and the série du volcan de Mahavel are grouped on the appellation "bouclier ancien" (ancient shield).

## Geochronology Of Piton de la Fournaise

In the massif of Piton de la Fournaise, the oldest lavas recognized are dated around 530,000 years. It demonstrates the simultaneous activity of Piton des Neiges and Piton de la Fournaise volcanoes over the last 500,000 years.

They have erupted similar basaltic lavas during a period of about 100,000 years between 530,000 and 435,000 years. The basaltic activity concentrates, primarily, in the zone of the Piton de la Fournaise, while the ultimate stage of activity of the Piton des Neiges consist of lavas belonging to a differentiated series ranging from hawaite to quartz-trachyte. This gives evidence to a shift-ing of the basaltic magma source south-eastward. The eruptive history of the Piton de la Fournaise is marked by three successive collapses situated by Potassium-Argon (K-Ar) and Uranium-Thorium (U-Th) dating between - 280,000 and -220,000 years for the first phase, separating phases 1 and 2 of the Piton de la Fournaise activity; between -43,000 and -19,000 for the sec-ond, which separates phases 2 and 3; the collapse of the Enclos Fouqué, the youngest phase, is attributed by the radiocarbon to the last 4,650 years. The duration of the identified phases of activity appears shorter and shorter, separated by slumps more and more frequent, and interestingly, more and more restricted volumes. This attests to a progressively increasing disequilibrium of the unstable south-eastern flank of the Island of Réunion, due to the shifting of activity.

## Structure and Dynamics Of The Central Zone of Piton de la Fournaise Volcano

The volcanic activity of Piton de la Fournaise is mostly concentrated on the central cone. The analysis of all the available geological data lead to the evidence and to describe a shallow reservoir complex beneath the summit area.

The presence of this complex in the past may be inferred from a surface volcano-tectonic feature; the imprint of a series of collapse-and-refilling episodes of the summit area is obvious in the topography. The centres of these paleo-collapse craters have drifted over a restricted area, a few hundreds of metres, with a pronounced east-west elongation that may reflect a slight tendency to an eastward migration. The present main summit crater, Dolomieu crater, resulted from an episode of collapse that took place in the early 1930s, following a noticeable inflation of the whole summit zone since 1927, and the eruption of a large volume of lava (135 million cubic metres ($4.77 \times 10^9$ ft$^3$) along a fissure opened along the northeast rift zone in the caldera. The deflation of the summit and the formation of a small pit crater inside Dolomieu, during the eruption of March 1986 at low altitude along the southeast rift zone, have recently provided further intangible proof of the existence of magma storage at a shallow depth beneath the summit.

This model is fully supported by the available structural geophysical data. Aeromagnetic data suggest the presence of a weakly magnetized volume of rocks beneath the summit, a fact that can be attributed to the presence of rocks above Curie temperature and of hydrothermally altered rocks. The same geological section, in addition to the presence of water impounded in the intrusion network, would fit with the low resistivities revealed by acoustic soundings. The very large (1.25 volts) spontaneous polarization (SP) anomaly of the summit can be associated to hydrothermal convection cells surrounding the reservoir complex.

With the data collected by the volcanic observatory (Obervatoire Volcanologique du Piton de la Fournaise) since 1980, it has been possible to refine the model of the reservoir for both geometrical and dynamical aspects. From the pattern of seismicity and deformation, it can be inferred that the storage complex is composed of discrete pockets separated by rigid walls. These storage units can be permanently or intermittently connected. They are distributed over a surface that corresponds approximately to that of Dolomieu crater. The shallowest ones can be located at only a few hundreds of metres the surface, but the main storage zone lies between 1 kilometre (3281 feet) and at least 2.5 kilometres (8202 feet). There are no definitive arguments to characterise below 2.5 kilometres (8202 feet).

Magma transfers between the deep zones and the shallow reservoir is thought to be discontinuous or irregular. The larger episodes of transfer may be associated to the major oceanite eruptions, the last being in 1977. Petrologically, the lavas erupted during the recent years are supposed to be trapped from batches of magma residing in the shallow reservoir complex.

## A Model For The Mechanical Behaviour And The Internal Structure Of Piton de la Fournaise

A tectonic model of the upper part of Piton de la Fournaise (the first 5 kilometres (16,404 feet)) is proposed on the basis of a theoretical mechanical investigation of the volcano. First, the geological model gives the broad but essential outline of the internal structure of the volcano in order to perform a second step, a numerical analysis of the stress field. The model includes a magma chamber located 3 kilometres (9842 feet) below the surface, surmounted by a central column representing a complex system of dykes and sills.

The stress field simulation has been done by geophysicists from the Institut de Physique du Globe de Paris (IPGP) using finite elements method on an axis-symmetric analysis in the static domain. The medium is elastic, continuous but heterogeneous, i.e., a specific temperature field has been imposed to the system and elastic parameters are temperature dependent. In addition to a normal pressure loading viz. body forces, hydrostatic, magmatic and external, thermal stresses were applied to the model.

Thirdly, the results viz. stress, distribution, magnitude and trajectory, are all interpreted in terms of mechanical failure. The final model is an ideal tectonic pattern for Piton de la Fournaise that comprises radial fractures, conical cracks with inward dipping, concentric cracks with outward dipping, etc.. Still there are many geological and geophysical evidences that will have to be confronted and rectified with the present model in order to find degrees of agreement or departure.

## Hydrogeological Approach Of Piton de la Fournaise

After examination of the synthetic overview of climatic, hydrological and geological environment of the *Massif du Piton de la Fournaise*, the hydrodynamic characteristics of the different structures and formations, lead one to the proposition of the classification of aquifers. The regional hydrogeology is determined using representative examples of the two large rivers, Rivière des Remparts and Rivière Lagevin. The perched sources of Plaine des Gregues and coastal emergences were prospected by aerial infra-red thermography. The result is a cartography of the emergence areas of the aquifers that underlines the links with the major structures affecting the volcanic edifice of Piton de la Fournaise.

A chemical and isotopic study of the precipitation and of the springs on the whole massif has allowed the determination of the approximate altitude of infiltration into the aquifers and to estimate the transit time of ground waters. On the other hand, the preliminary first results do not allow one to have been recorded with respects to the temperatures, conductivities and stable isotopes of some waters entering the tunnel of Rivière de l'Est.

## Eruption Characteristics Of Piton de la Fournaise

There are many aspects to examine when considering the eruption characteristics of Piton de la Fournaise. Those aspects include the location of the eruption, the lava situation, types of damage sustained, specific type of volcanic activity, and the Volcano Explosive Index (VEI).

Figure 3–1 through 3–5, in conjunction with the following guideline key, will serve to consolidate the graphic summarization of data for the 171 eruptions of Piton de la Fournaise.

### ERUPTION LOCATION
C = Central crater eruption
E = Eccentric (parasitic) crater
R = Radial fissure eruption
F = Regional fissure eruption

### DAMAGE TYPES
F = Fatalities
D= Destruction of land, property
M= Mudflows (lahars)
T = Tsunami (giant sea waves)

### ACTIVITY TYPES
E = Explosive (normal)
N = Nuée ardentes, pyroclastic flows
P = Phreatic explosions
S = Solfataric activity

### LAVA SITUATION
F = Lava flow(s)
L = Lava lake eruption
D = Dome extrusion
S = Spine extrusion

VEI = Volcanic Explosive Index (See Appendix B)

| <u>C</u> | <u>E</u> | <u>R</u> | <u>F</u> |
|---|---|---|---|
| 41 | 40 | 57 | 0 |

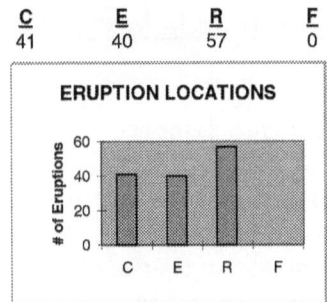

Figure 3-1. Eruption Locations

| <u>F</u> | <u>L</u> | <u>D</u> | <u>S</u> |
|---|---|---|---|
| 143 | 1 | 2 | 0 |

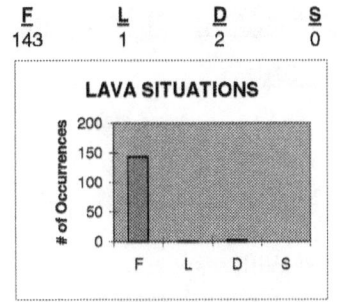

Figure 3-2. Lava Situations

| <u>F</u> | <u>D</u> | <u>M</u> | <u>T</u> |
|---|---|---|---|
| 0 | 1 | 1 | 0 |

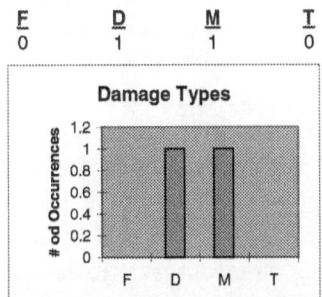

Figure 3-3. Damage Types

| <u>E</u> | <u>N</u> | <u>P</u> | <u>S</u> |
|---|---|---|---|
| 81 | 0 | 0 | 0 |

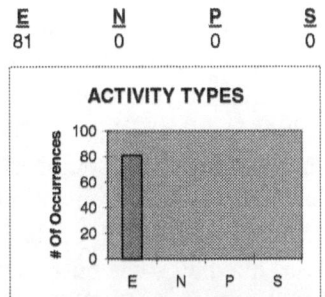

Figure 3-4. Activity Types

| VEI Index | <u>0</u> | <u>1</u> | <u>2</u> |
|---|---|---|---|
| Count | 41 | 12 | 98 |
| % | 0.27 | 0.08 | 0.65 |

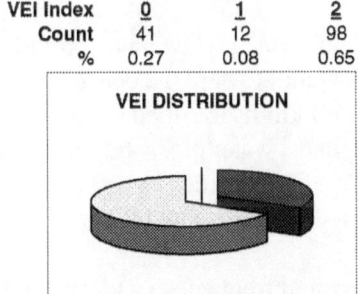

Figure 3-5. VEI Distribution

# Assessment Of Natural Hazards At Piton de la Fournaise Volcano

As it has been previously stated, Piton de la Fournaise, is the active shield volcano of Réunion Island in the western Indian Ocean. It is one of the most regularly active volcanoes in the world, averaging one event every 10 months. with a mean effusion rate calculated at 0.3 m/s (.98 ft/s) during the last 60 years. With respect to natural hazard assessment it is, of course, of great concern to the inhabitants of Réunion Island regarding the on-going and potential of the volcano.

The flanks of Piton de la Fournaise volcano are sparsely inhabited by scattered rural, cosmopolitan and multireligious populations whose behaviour in front of eruptions is as yet not really well known. This population is stretched over on both sides of the littoral road in two major linespread villages, Pointe du Tremblet and Bois Blanc (9,000 inhabitants) in the vicinity of or on the spots of the northeast and southeast active lateral rift zones outside the caldera, where 5% of the eruptions occur. None of the population, of course, live inside the caldera where 95% of the eruptions occur. It is only crossed by the road and occupied by some bush plantations growing on recent lava flows and scoria's. A small town, St. Philippe (24,000 inhabitants), lies far from these active zones. No industry, no regional interest building, e.g., hospitals, etc., are set up on these volcano flanks.

The principle type of eruption of Piton de la Fournaise is "Hawaiian" providing mainly fluid lava of oceanic basalts running downhill toward the ocean at a rate of approximately 400 m/day (1312 ft/day) to 60 km/hr (37 mph). Sugar cane and vanilla plantations, forests, roads, electric lines, water pipes, buildings and houses are episodically destroyed by lava flows. Creeks filled by lavas change their bed and surface runoff with some major risks for the population.

Lava fountains may produce Pelee's hairs (decenial frequency) widespread by tradewinds over the immediate surrounding areas (115,000 inhabitants concerned). Cattle and market gardens may be then respectively killed and/or spoiled. Phreatic eruptions produce cinders that may be scattered over the entire island such as with the eruptions of 1802, 1812, and 1861. Sulfur can be exhausted in significant quantities from vents (50 to 100 tons/day) during an eruption with consequences unknown.

In addition to the direct volcanic hazards, the general population has to reckon with the indirect hazards. For example, slumpings occur on the ancient collapsed structures walls, deeply eroded by rivers. They may be triggered by water charge (world record rainfalls between 12 hours to 15 days) or by rare regional seismics (1927). Landslides, dams and mudflows (lahars) also occur (1927 & 1962), determining significant risks for about 15,000 inhabitants. As at the Kilauea volcano (Nilina fault zone) in Hawaii, Piton de la Fournaise volcano is threatened by huge landslide blocks with the slumping of entire rift-zone-bounded block themselves, concerning about 4,000 persons.

An alert plan (plan ORSEC) is set at work for each eruption by the Chief Administrator supported by scientific survey provided by IPGP's Observatoire Volcanologique du Piton de la Fournaise. This includes seismic, deformation, and magnetic permanent networks. Additional support is provided by University of la Réunion (geology), and BRGM Service Géologique Régional (geology, computerized mapping, computer data banks, and landslides phenomena) and long range of event forecasting of Piton de la Fournaise is provided by the Southwest Volcano Research Institute in the United States. No evacuation plan or special permanent rules are in existence.

For natural hazards, a new research field in progress (BRGM) deals with volcanic and mud and lava flows simulation. Output from the computer is accurate to 1/25,000 lava flow mapping contours. BRGM was instrumental in the development of an original calculation programme to accomplish this task.

## Probabilistic And Statistical Characteristics Of The Historic Activity Of Volcano Piton de la Fournaise

The statistical and probabilistic approach to Piton de la Fournaise's volcanic activity for the last 150, 50, and 15 years is based on careful investigation of the historical archives. Historical periods of volcanic activity and repose have a statistical behaviour linked to a Poisson distribution law process. Further, Piton de la Fournaise's eruptions can also be linked to internal phenomena due to refilling of different magma reservoirs.

Three major volcanic cycles for the last 50 years have lasted between 15 and 30 years each. The main lava output for this last half century is 0.16 $m^3$/sec (5.6 $ft^3$/sec); but the lava outputs for the eruptions vary from 0.32 $m^3$/sec to 10 $m^3$/sec (11.3 to 353.1 $ft^3$/sec). While the internal processes (activity within the magma chambers) essentially determine the types of activity or repose of Piton de la Fournaise, it is the external processes, such as earth tides, due to the different moon phases, which have a determinant effect on the triggering of eruptions. The earth's tides are determinant within a two-week period due to the transfer of magma to the shallow reservoir of approximately 1–2 kilometres (3281–6562 feet) in depth.

## Piton de la Fournaise Volcanic Activity

As previously mentioned, Piton de la Fournaise is one of the most regularly active volcanoes in the world, averaging one event every 10 months. Moreover, its effusion rate has increased in the last 30 years to 0.3 m³/sec (10.6 ft³/sec). The oldest data on Piton de la Fournaise's historical activity given by sailors, date from 1644. Archive investigations have brought to light 171 events (see Appendix C).

95% of the historical activity occurred inside the caldera of Enclos (see Figure 3–6) which collapsed 2300 years ago; in the last 3½ centuries, only 8 eruptions or 5% of the activity occurred outside the Enclos (see Figure 3–7).

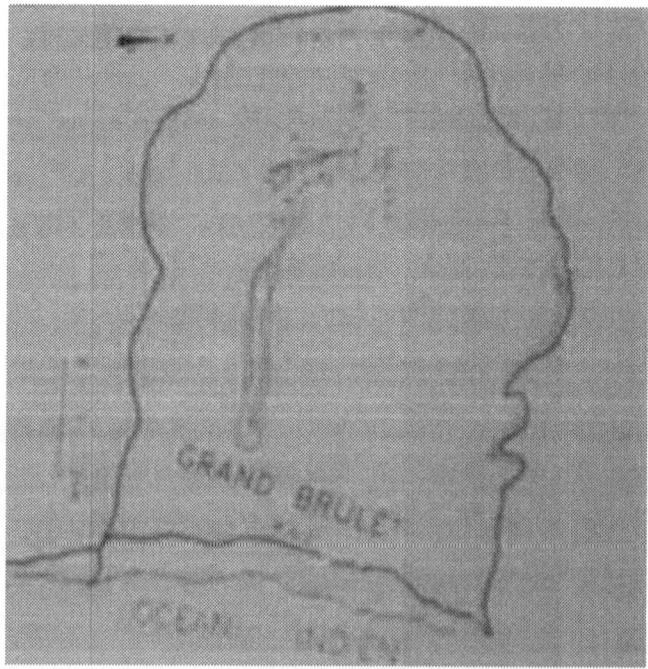

*Figure 3–6. The Layout Of Volcano Piton de la Fournaise*

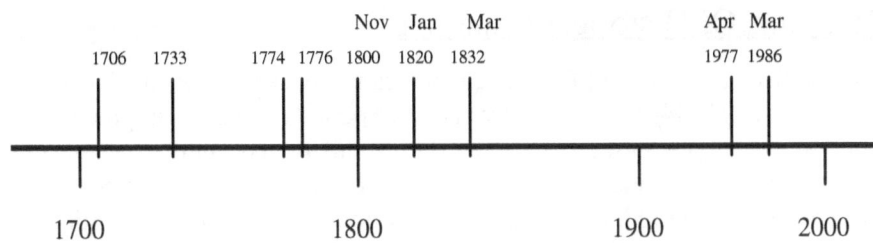

*Figure 3–7.  Historical Eruptions (8) Of Piton de la Fournaise Volcano Occurring Outside The Enclos Caldera*

The latest occurred in 1977 then again in 1986. Basaltic lava flowed through the villages of Piton Ste. Rose on the northern flank in 1977 then Le Tremblet on the southern flank in 1986. The lava flows destroyed forests, sugar cane and vanilla plantations, several houses, and covered the main road.

During the 19th century, approximately 16 huge lava flows reached the Indian Ocean, then only 6 flows of less importance reached it during the 20th century.

Although the number of known events is great (171), it is insufficient for forecasts of volcanic hazards. In terms of quality and precision of historical data, the only events taken into account where the events of the last 150 years (the 1844 to 1984 period), then those of the last 50 years (the 1931–1986 period), then those of the last 15 years (the 1972–1986 period). It should be noted that the data collected are not complete; numerous events, in particular shorter ones, were not recorded before 1980. (See Appendix A for selected views of Piton de la Fournasie's eruptive activity.)

## Periods Of Repose

For the last 150 years (1844–1985), durations of intervals recorded between eruptions (repose periods) vary from 2 weeks to 8 years. These data are not significant, however, because many eruptions were not recorded. If one considers contemporary chronicles with more data, one can find some indications for the last 50 years (since 1931) and for the last 15 years (since 1972) (see Table 3–1). One considers that a repose period of inactivity begins just after the triggering of an eruption, and lasts until the beginning of the next eruption.

## Durations Of Historical Eruptions

For a statistical approach (Table 3–2), one can only use the events during the last 150 years (1844–1985), data from earlier periods being too imprecise and with too many gaps in the data. Even during this period, numerous volcanic events, mostly short ones, were not observed or recorded. One can conclude, however for this period, the duration of eruptions is generally short (Stieltjes, 1984). i.e., approximately 25% last 1 to 2 days, 39% last 1 week or less, 75% last 1 month or less, and 85% last 2 months or less. Short lived eruptions (1 to 2 days) are probably much more numerous.

The duration of eruptions has increased in the last 50 years since 1931. For the last 15 years (1972–1985), short eruptions (1 or 2 days) and medium eruptions (1 to 3 weeks) are more numerous and there is no known long eruptions (greater than 1 month) for this period.

## Relation Between Durations Of Eruptions And Periods Of Repose Of Piton de la Fournaise Volcano

Except for very short events, the duration of an eruption has a direct influence on the duration of the following periods of repose. After eruptions of less than 10 days, the period of repose lasts 5.5 months (166 days) on average, but most short periods of repose (less than 2 months) follow eruptions lasting less than 1 month. In four cases out of six, long periods of repose (more than 8 months), follow a long eruption (more than 46 days), and in one case, a short eruption (less than 10 days).

On the other hand, the duration of period of repose does not influence the duration of following eruptions; after very short repose (less than 60 days), there are no flash eruptions (1 to 2 days). Sixty percent of the eruptions are of medium length (1 to 3 weeks), and 40% are long eruptions (more than a month). After long repose (more than a year), 25% are flash eruptions, 45% are medium length eruptions and 30% are long eruptions.

Thus long eruptions do not necessarily result from long periods of repose of Piton de la Fournaise. The duration and types of eruptions of the Piton de la Fournaise volcano are much more influenced by factors other than the duration of volcanic repose, namely the location of an eruption in a volcanic cycle and the volumes of lava previously emitted.

# Eruptive Cycles And Volumes Of Magma Emitted

*Eruptive Cycles*

One can distinguish 3 major cycles for the last 50 years of Piton de la Fournaise eruptions. Each cycle begins with a huge eruption and emission of a large volume of oceantite magma. The 1931 eruption marked the beginning of a 30-year cycle (1931–1960), the 1961 eruption was followed by a 15-year cycle (1961–1976), and another cycle began with the eruption of 1977.

One can link these decade-long cycles with the time for refilling the magma reservoir at an intermediate depth of 5 to 10 kilometres (3.1 to 6.2 miles) from the upper mantle at 20 to 30 kilometres (12.4 to 18.6 miles). This intermediate reservoir generally feeds short eruptions on the flanks of the dome and inside the two summit craters, Dolomieu and Bory. Large eruptions at the beginning of a cycle occur both inside and outside the Enclos caldera, but all produce oceanic magma from deep reservoirs.

*Table 3–1*  *Statistics of Repose Periods at Piton de la Fournaise Volcano For Historical Times, Over The Last 150, 50 And 15 Years.*

| Periods of Repose | 1844-1986 Period (Last 150 years) | | 1931-1986 Period (Last 50 years) | | 1972-1986 Period (Last 15 years) | | Remarks |
|---|---|---|---|---|---|---|---|
| | Mean Duration | % Of Events | Mean Duration | % Of Events | Mean Duration | % Of Events | |
| Short Repose | | | 2 months (<55 days) | 26% | 2 months (<53 days) | 70% | The daily eruptive probability is constant at 0.0.45 (1/22) |
| Mean length Repose | Too many gaps in the recording of eruptions to give any significance to a study of intervals between eruptions. | | 3-8 months (94-242 days) | 42% | 1 year (396 days) | 25% | During the 110 days of repose, an eruption will be"statistically impossible (time for refilling the intermediate depth magma reservoir). After 110 days the daily eruptive probability is 0.045, then slowly decreases (0.0025 after 850 days). |
| Long repose | | | > 1 year (>349 days) | 32% | > 1 year | 5% | |

*Table 3–2*     *Statistics Of Duration Of Eruptions at Piton de la Fournaise Volcano Over The Last 150, 50 And 15 Years.*

| Eruptive Periods | 1844-1986 Period (Last 150 years) | | 1931-1986 Period (Last 50 years) | | 1972-1986 Period (Last 15 years) | | Remarks |
|---|---|---|---|---|---|---|---|
| | Mean Duration | % Of Events | Mean Duration | % Of Events | Mean Duration | % Of Events | |
| Short Eruptions | 1-2 days | 25% | 1-2 days | 20% | 1-2 days | 30% | |
| Middle Length | 4-26 days | 50% | 4-26 days | 50% | 3-21 days | 70% | The daily Eruptions probability for an eruption has a linear increase from 0 at the beginning of an eruption. Few of the eruptions last less than 1 week and more have a duration close to the periodicity (20 days) |
| Long Eruptions | > 30 days | 25% | > 30 days | 30% | -- | -- | |

## Volumes Of Emitted Magma

Within the 3 latest eruptive cycles, one may notice irregularities in the output of lava since 1931. For the period 1931–1960, the mean output was low: 0.16 $m^3$/sec (5.7 $ft^3$/sec). It has increased for the recent period from 1960 to 1985 to 0.78 $m^3$/sec (27.5 $ft^3$/sec). For this whole period (1931–1985), the mean lava output of Piton de la Fournaise is 0.3 $m^3$/sec (10.6 $ft^3$/sec). It is generally lower than the mean output of lava from Kilauea and similar to that of Mt. Etna in Sicily.

It seems that there is a low probability that the output line will go outside of the variation band. So after the 1977 eruptions, one could expect a period of rather small eruptions. In 1986, the probability of having an important eruption was still low.

With a 0.16 m$^3$/sec (5.7 ft$^3$/sec) mean output of lava, the rate remains inside a narrower band; this increases the precision of eruptive probability (but with less certitude). As a matter of record, the maximum volume emitted during an eruption inside a cycle is 30 million m$^3$ (10.6 X 10$^8$ ft$^3$) for the first eruption (1931), or this volume may be spread over on several eruptions from 1961 to 1966.

Everything about the behaviour suggests that repose periods of Piton de la Fournaise volcano are related to the time needed to recharge a middle depth reservoir (5–10 kilometres (3.1–6.2 miles)) with a volume close to 30 million m$^3$ (10.6 X 10$^8$ ft$^3$) coming from a deeper reservoir (20–30 kilometres (12.4–18.6 miles)) which may erupt outside the central emission zone inside the caldera: huge volumes of oceantite are so produced from the peripheral emissive zones (Grandes Pentes, external ridges of the Enclos: branches of Ste. Rose and Ste. Philippe).

During the 1931 and 1977 eruptions, the superficial magma chamber may not have played its role of buffer reservoir for oceantite coming directly from the upper mantle and erupting far from the central dome and main rift zone inside the caldera. During the 1961–1966 period, however, at the beginning of a 15-year cycle, the intermediate depth reservoir seems to have played a weak role as a buffer; instead of only one large eruption beginning the cycle, as in the 1931 eruption, the emission of the huge volume of new magma extended over a five-year period.

## Relationship Between The Volume Of Magma And The Duration Of Activity And Repose Periods

*Volume Of Magma Related To Duration Of Eruptions*
There is a relationship between the duration of eruptions and the volumes of magma erupted since 1972. Four types of eruptions can be distinguished from observing and analysis of Table 3–2:

- *very short and weak eruptions*, emitting less than 0.1 million m$^3$ (3.5 X 10$^6$ ft$^3$) and lasting less than 1 day;

- *short but vigorous eruptions,* lasting 1 to 2 days and producing about 10 m$^3$/sec (353.1 ft$^3$/sec) of lava. Only 3 volcanic events of that type have occurred since 1972;

- *medium length eruptions,* lasting 1 week to 1 month and producing about 2.3 m³/sec (81.2 ft³/sec) of lava. These represent about 40% of the eruptions since 1972;

- *long but weak eruptions,* lasting more than 1 month.

*Volume Of Magma Related To Duration Of Repose*

The volume of magma emitted during an eruption is not a good clue for the duration of the next repose period. After the emission of weak volumes of lava (less than 1 million m³ (353.1 X 10⁷ ft³)) the following repose intervals can be divided into three types. Approximately 40% are short repose periods of less than 2 months, 30% last two to eight months, and 30% last more than 8 months.

After emissions of large volumes of lava (more than 16 million m³ (5.6 X 10⁸ ft³)), no repose periods last less than 5 months, about 20% of the repose periods last 5–8 months, and about 80% are longer than 8 months. Of the latter about 40% continue more than 2 years and 18% more than 3 year. As an example, the eruptions of 1981 and 1983, both of which lasted 1 month, produced 8 million m³ (2.82 X 10⁸ ft³) of lava each. The first was followed by a 977 day repose period and the second by one of 46 days.

On the other hand, the volume of magma emitted is to the duration of previous repose period. After a repose period of less than 3 months, about 10% of the eruptions emit less than 1 million m³ (3.53 X 10⁷ ft³) 70% produce 1–10 million m³ (3.53 X 10⁷ to 3.53 X 10⁹ ft³), and 20% erupt 10–100 million m³ (3.53 X 10⁸ to 3.53 X 10¹⁰ ft³) of lava. After a repose period of more than three months, about 40% of the eruptions emit less than 1 million m³ (3.53 X 10⁷ ft³), 30% erupt 1 to 10 million m³ (3.53 X 10⁷ to 3.53 X 10⁹ ft³), and 30% erupt 10 to 100 million m³ (3.53 X 10⁸ to 3.53 X 10¹⁰ ft³). In conclusion, it can be said that the duration of a repose period is linked to both the duration and volume of magma of the previous eruption.

# Influence Of Earth Tides On The Triggering Of Eruptions

The effect of earth tides on triggering of volcanic eruptions has been demonstrated in numerous cases (e.g., Mauk and Johnston, 1973; Hamilton, 1973; Dzurisin, 1980; Martin and Rose, 1981. The attraction of the sun on the surface of the earth varies during the year, with its lower value at the winter solstice, but it is always less than that of the moon and can be considered, for all intents and purposes, negligible. Three lunar cycles influence the earth's tides:

- the *synodic cycle* lasting 29.5 days when the moon is in conjunction or in opposition to the sun;

- the *anomalous cycle* lasting 27.6 days and representing the variation of the distance from the earth to the moon. At its perigee, the moon is closest to the earth; at its apogee, it is at its maximum distance from the earth;

- the *tropical cycle* lasting 27.3 days represents the variation of the declination of the moon's orbit.

These three cycles are independent and interact to produce many kinds of phase displacements. During 1821, for example, the lunar perigee of maximum declination (28.5°) of the new moon coincided with the summer solstice. At that time, three eruptions occurred on the central dome of Piton de la Fournaise and lava flows reached the sea.

The timing of the eruptions of Piton de la Fournaise through these three types of lunar cycles shows that the lunar apogee is much more likely to trigger an eruption than the perigee. The lunar perigee is favourable for triggering eruptions only when there is a conjunction with the full moon that occurs every 7 months.

The phase displacements between the different cycles may also be significant. The configuration of a new moon at the perigee, following 14 days later by that of a full moon at the apogee is very favourable for triggering an eruption of Piton de la Fournaise. About 40% of its eruptions have occurred within 3.5 days of this phase. When the moon is at the apogee, it is more propitious too than a full moon at the perigee, which has a probability of only 5% of triggering an eruption.

The same phase configuration comes back every seven months. Many repose periods last about seven months; this relation separates the medium length periods of repose into two classes; those lasting less than eight months and those lasting more than 1 year.

# The 1992 Eruption Of Piton de la Fournaise

Volcano Piton de la Fournaise erupted last on August 28, 1992 after an erratic increase in seismic activity for the prior seven months. Several seismic episodes were recorded prior to the actual eruption with the calculated epicentres mainly located beneath crater Dolomieu, with the foci distributed between the summit and sea level. This type of seismic behaviour is suggestive of a rather shallow magma pocket and, in this case, giving rise to the notion that a magmatic intrusion is taking place within Dolomieu. On both distant and local recorders, vertical ground deformation was duly noted on all stations with a 130 microradian maximum displacement noted.

Outpouring occurred from a fissure within the Dolomieu crater and showed an approximately 40 metre lava fountain and emitted a small lava flow. The fissure rapidly propagated southward, crossing the southwest rim of crater Dolomieu, and produced a very short lava flow (flow # 1) on the slope (see Figure 3–6). Successive openings of eruptive vents produced 4 successive lava flows (flows 2 to 5) (see Figure 3–6). Activity had been early concentrated on the southeastern slope of crater Dolomieu where the high lava fountains (40 metres (131 feet)) were built up and eruptive cone, Zoé.

Most flows were of short duration, about 3 hours maximum. Very strong degassing was also noted and high velocities of flow # 5 was noticed about 3 hours after the start of the eruption.

Direct observations made on August 28th determined the lava fountain height at 40 metres (131 feet), and the dense degassing consisting of $SiO_2$-rich gases. The velocity of the lava was estimated at the foot of Zoé at approximately 4 metres/sec (13.1 feet/sec). Lava flowed downward on the slope of the Grand Brulé and covered an area estimated to be 1 million $m^2$ (10.8 million $ft^2$). With a mean width assumed for flow # 5, it is calculated that the Zoé crater emitted a volume of about 5 million $m^3$ (1.77 X $10^8$ $ft^3$) of lava. The same calculations for flows # 1 through # 4, using a mean thickness of 2 metres (6.6 feet) for the flows on the slopes of Dolomieu yield a volume of approximately 500,000 $m^3$ (1.8 X $10^7$ $ft^3$). Therefore, for the August 1992 eruption, a total of approximately 5.5 million $m^3$ (1.9 X $10^8$ $ft^3$) of aphyric basalt was emitted.

# Conclusions About Piton de la Fournaise

This statistical and probabilistic approach to the activity of Piton de la Fournaise volcano for the last 150, 50 and 15 years becomes more meaningful for more recent periods because of the more accurate data and records now available. The relations indicate that the eruptive processes are related to internal and external phenomena.

(1) For the last 2 centuries, the activity of Piton de la Fournaise has been quite regular with a mean repose period of 10 months. In fact, the longest inactive period recorded in the last 15 years was about 2.5 years; the shortest was about 2 weeks.

(2) Most eruptions of Piton de la Fournaise last less than one month (75%), and many (39%) last less than one week. About 25% of the eruptions last less than two days.

(3) Periods of volcanic activity and repose have a statistical behaviour linked to a "Poisson Distribution" law process. This means that the activity is not random but is linked to an internal process.

(4) The duration of an eruption has a direct influence on the length of the following repose periods. But the contrary is not true; the length of a repose period does not influence the duration of the following eruptions.

(5) Three major volcanic cycles can be distinguished during the last 50 years, lasting between 15 and 30 years each. The third is still in progress. Each cycle begins with a huge outpouring of oceantite magma, generally flowing down to the Indian Ocean. Eruptions within a cycle rarely reach the ocean. The mean lava output for the last 50 years is 0.16 $m^3$/sec (5.7 $ft^3$/sec). Three groups of eruptions are characterised by their average outputs of lava: 10 $m^3$/sec (353.1 $ft^3$/sec), 2.3 $m^3$/sec (81.2 $ft^3$/sec), and 0.32$m^3$/sec (11.3 $ft^3$/sec).

(6) Relationships between the volume of magma erupted and the duration of active and repose periods are very clear for Piton de la Fournaise. The duration of a repose period is linked both with the duration of the previous eruption and with the volume of magma emitted during the previous eruption. The behaviour suggests that repose periods of Piton de la Fournaise volcano are related to the time needed to recharge an intermediate reservoir at a depth of 5 to 10 kilometres (3.1–6.2 miles)

and having a volume close to 30 million m$^3$ (10.5 X 10$^8$ ft$^3$). This reservoir is fed by a deeper reservoir 20 to 30 kilometres (12.4 to 18.6 miles) below the volcano. The intermediate depth reservoir normally acts as a buffer, but it sometimes may be overfed by a flow of magma from a deeper source at 20 to 30 kilometres (12.4 to 18.6 miles), which erupts from the central zone inside the caldera. At these times, huge volumes of oceantite are produced outside the customary ridges zones.

(7) While the internal volcanic processes and activity in the magma chambers determine the duration and types of activity or inactivity of the Piton de la Fournaise volcano, it is the external processes, mainly earth tides, due to the different phases of the moon, which have a determinant effect on triggering eruptions.

The attraction of the moon adds its effects to tension due to swelling of the external layers of the shield, giving the tensional stresses that provoke the opening of fissures and eruptions. The maximum amplitude of the combined forces of the moon and sun is reached every 29.5 days. The sun's attraction is negligible compared to that of the moon and does not itself trigger eruptions. Moreover, attraction of the planets, even when they are in conjunction, is negligible compared to that of the moon.

Earth tides are effective within a two week period, owing to the transfer of magma toward the superficial reservoir, 1 to 2 kilometres (.62 to 1.2 miles) below the volcano.

# CHAPTER 4

# THE VOLCANO OBSERVATORY

## Formation Of The L'Observatoire du Volcanologique du Piton de la Fournaise

L'Observatoire du Volcanologique du Piton de la Fournaise, is the volcanic observatory of Réunion Island and was established in the summer of 1980 by the Institut de Physique du Globe de Paris (IPGP). L'Observatoire du Volcanologique du Piton de la Fournaise operates under the auspices of and in conjunction with the business assurance and support of le départment de la Réunion. The IPGP is fundamentally a research facility within the domain of global geophysics including; magnetism, gravity measurements, seismology, volcanology, geochemistry, geophysics, and tectonic plate movement. IPGP also has interests in the exterior environments as well as the interior of the earth environs.

L'Observatoire du Volcanologique du Piton de la Fournaise is intentionally situated near the main caldera of Piton de la Fournaise (L'Enclos Fouqué) and is located only 15 kilometres (9.3 miles) from the main volcano caldera. Since its creation in 1980, L'Observatoire du Volcanologique du Piton de la Fournaise has maintained constant surveillance of Piton de la Fournaise on a 24 hour, 365 days per year basis. The necessity for the creation of L'Observatoire du Volcanologique du Piton de la Fournaise was primarily prompted by the inhabitants of Réunion Island, who are, understandably, concerned about their safety and welfare of their person and property on the island. In 1986, for example, and for the second time since 1800, large flows of lava have traversed down the slopes of Piton de la Fournaise and into the sea (Indian Ocean). On the way, damage has been sustained to nearby villages, e.g., Bois Blanc, St. Phillipe, Ste. Rose, Pt. Tremblet, etc., but there has been no deaths, attributed to date, as a result of any eruption of Piton de la Fournaise.

There is a resident staff at the observatory consisting of a Director who operates under the direction of IPGP's Jean-Louis Cheminée, Director of all volcano observatories, a chief seismologist, a cartographer, several technicians (electronic, instrumentation, etc.), a secretary and necessary support staff, hired from the local populace. There is almost always a constant steam of visitors to L'Observatoire du Volcanologique du Piton de la Fournaise, mostly from France's IPGP or universities around and about France and in some cases, other universities. Most visitors are conducting specific scientific research projects within their discipline and some are working on their Masters or Doctorates in their chosen field. There are adequate boarding facilities located at L'Observatoire du Volcanologique du Piton de la Fournaise to accommodate several visitors.

## Monitoring Of The Volcano By L'Observatoire du Volcanologique du Piton de la Fournaise

There are several types of scientific monitoring that keep the volcano under surveillance at L'Observatoire du Volcanologique du Piton de la Fournaise. They include instrumentation to monitor seismic, tilt, deformation, gravity, distance measurement, and fissure width.

*Seismic Monitoring*

The monitoring and detection of seismic events under the mountain is paramount in the forecasting and detection of a likely eruptive event of Piton de la Fournaise. There are currently 15 seismic monitors about the island used to monitor the seismic activity of Piton de la Fournaise. There are four seismometers within the l'Enclos Fouqué and the other monitors are outside the main crater area (see Figure 4–1). The seismic measuring equipment is powered by solar panels located near each seismometer each with their own electronics package. The collected information is radio relayed back down to the volcano observatory monitoring equipment (see Plates 23–36). The monitoring is performed 24 hours a day for every day of the year. Albeit that the observatory is not manned 24 hours a day, (except during an event) there is a phone patch system installed in the observatory. If there are 3 seismic events detected within a five minute period, the telephones at the residences of the Director of L'Observatoire du Volcanologique du Piton de la Fournaise, the principal seismologist and other principals, are immediately

rung with a special ring to alert observatory personnel of a possible significant event. Figure 4–2 illustrates the quiescent volcano with two 8 seconds
shocks, detected by two seismometers, just prior to the 15th of April 1990
eruption of Piton de la Fournaise. Figure 4–3 illustrates a 6 second event
detected by three seismometers just prior to the eruption on the 15th of April
and Figure 4–4 shows the full eruption of the volcano.

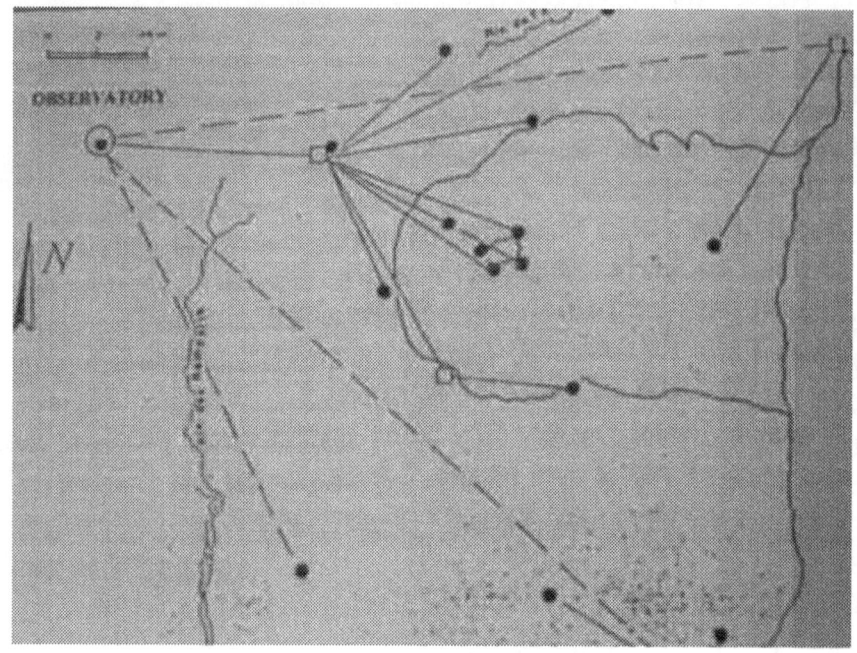

*Figure 4–1. Seismic station locations about Réunion Island.*

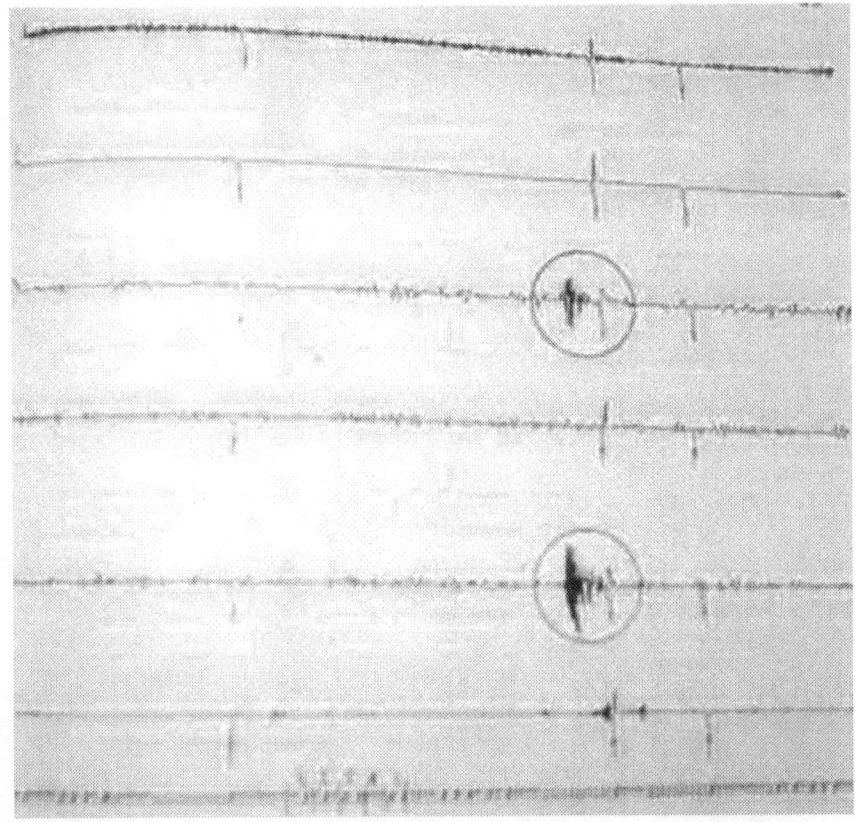

Figure 4–2.    The seismograph showing the quiescent state of Piton
de la Fournaise and two 8-second shocks just prior to
the April 15, 1990 eruption.

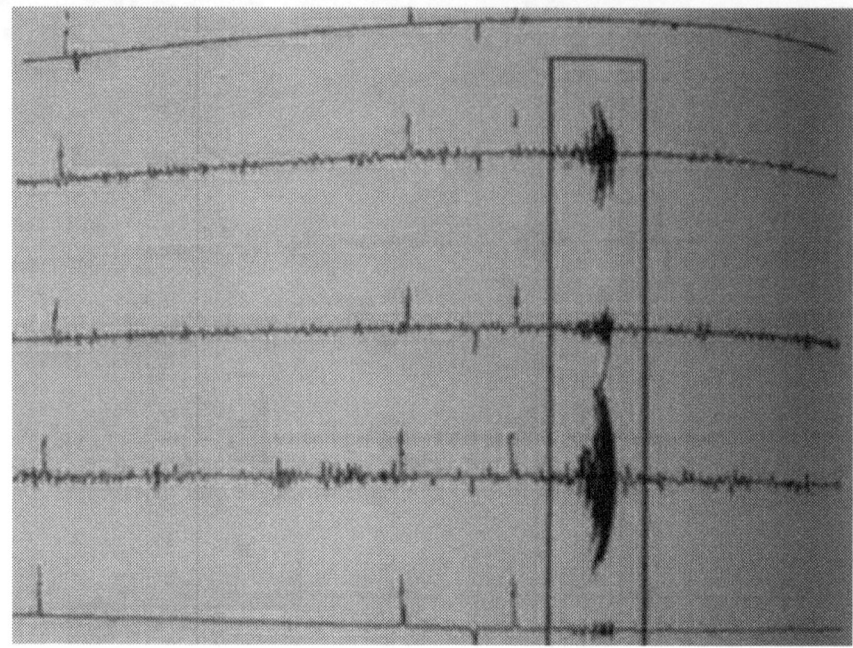

Figure 4–3.   *The seismograph showing three 6-second shocks just
prior to the April 15, 1990 eruption of Piton de la
Fournaise.*

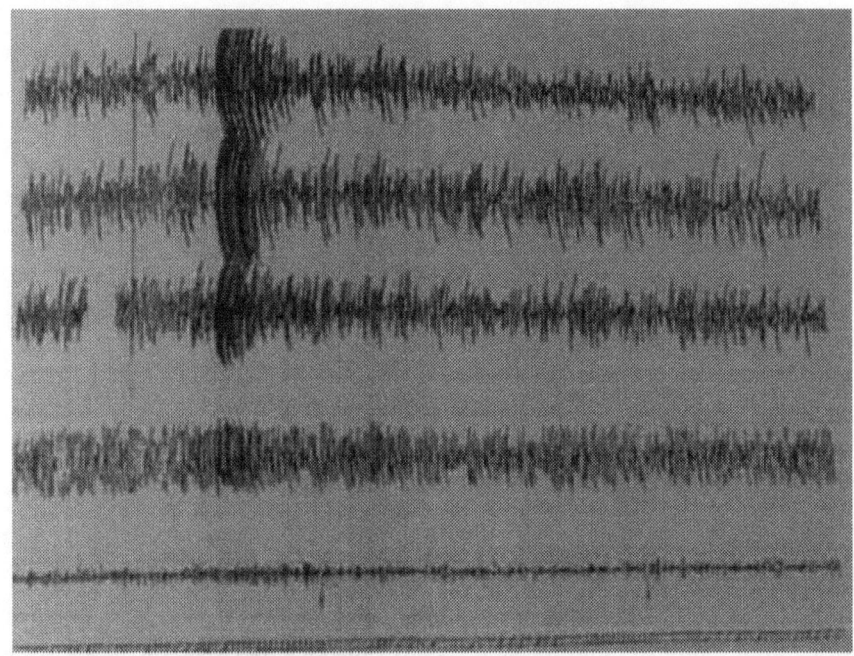

*Figure 4–4.    Seismograph of Piton de la Fournaise in full eruption*
*on April 15, 1990.*

The seismic equipment is held in constant repair and in good order by the electronic technicians who maintain all seismic and other equipment used in the monitoring of the volcano (see photo Plates in Appendix A.)

*Other Volcano Monitoring Apparatus*

There are principally four other types of monitoring of Piton de la Fournaise that takes place on a 24 hour per day, 365 days per year basis. They are the monitoring of deformation, gravity, inclination, and fissure width.

Deformation is detected by measuring the distance between two points with an argon (Ar) laser. This measurement is repeated every 5 minutes and relayed back to the observatory monitoring equipment. There is currently one laser in place at Piton de la Fournaise. If the volcano shows a significant change of several centimetres in the distance between the two points, the volcanologists at the observatory are notified by the monitoring equipment setting off an alarm.

Deformation is also detected with the aid of several (7) inclinometers set about the main crater area of Piton de la Fournaise. Surface measurements are made and monitored continuously from these 7 inclinometers and each is capable of measuring inclinations (tilt) as little as 1 microradian. Inclination will occur if there is a magma intrusion from under the mountain causing the surface area to swell and incline from its normal state. Again, via radio relay, if there is a detection of swelling, the volcanologists at the observatory are notified.

Gravity measurements are made with magnetometers of which there are eight scattered about the volcano. An intrusion of magma from under the volcano would be detected as a change in gravity due to the influx of a great mass of magma. This change in gravity is then detected by the magnetometer and radio relayed back to the volcanic observatory for analysis.

Fissure width is also monitored about the main crater area with the use of an extensometer on those fissures significant enough to warrant monitoring. The extensometer is placed within a particular selected fissure. The extensometer permits the monitoring and measurement of distance between the sides of the fissure. The slightest change in the width of the fissure (signifying movement of the mountain) will be radio relayed back to the observatory.

*Other Monitoring Apparatus*

There is also other types of monitoring that is performed by the observatory staff. There is a meteorological station on the observatory site proper that measures daily temperature, humidity, rainfall, etc.. This information, along with the normal volcanic monitoring data from Piton de la Fournaise, is duly recorded and sent via satellite, the Aerial Reconnaissance Global Orbiting Satellite (ARGOS) system to a worldwide volcano information monitoring network (Ortega, 1993).

The volcanic observatory at Réunion Island is also acting as a tsunami (tidal wave) warning centre of sorts. A seismometer is located out in the Indian Ocean and is primarily used for the detection of undersea earthquakes which could generate tsunamis. This Indian Ocean seismic information is recorded on the seismograph located within the observatory (see Appendix A). This information, in turn is the transmitted to IPGP and other points via the ARGOS satellite system.

*The Monitoring Equipment Bay*

Located within the volcano observatory are several instruments which are receiving transmitted information from the volcano transmitters. The instrumentation in the observatory constitutes the heart of the observatory. (see Appendix A) Seismic information telemetered down from the volcano to the observatory is fed into amplifiers where the received signals are further processed. Processed seismic signals are then sent to the strip pen recorder chart, are monitored by the special telephone alert system, and are recorded on magnetic tape recorders. Additional equipment includes the tsunami seismic warning signals as monitored in the Indian Ocean, which are also recorded on drum pen recorders, as well as time signal standards equipment.

# CHAPTER 5

## COMPUTER SOFTWARE PACKAGE *"ERUPTION"*

An experimental computer programme, specifically designed for the MS-DOS based PC computers (Trombley, 1990), in an attempt to forecast volcanic eruptions, has been developed and tested over the past six years. This new application programme uses the fundamental concept of the Poisson distribution paralleling the works of Wickman (1966) and De La Cruz-Reyna (1991).

The difference lies in the application of dynamic Poisson distribution parameters in the programme as well as the conventional parameters of the Poisson distribution. As an eruption occurs, the database is immediately updated in *"ERUPTION"* or, if there is no eruption during the current year, the Poisson parameter, $\mu$, the calculated average eruption rate per year, is updated as each year passes. Currently, post-Holocene data for 448 strato-volcanoes or near strato-volcanoes about the world has been loaded into *"ERUPTION"* and analyzed. Data of shield volcanoes has not been loaded as shield volcanoes typically do not exhibit Poisson distribution characteristics. Similarly, submarine volcanoes are also omitted. Data has been retrieved from the archives of the Smithsonian Institution in Washington, D. C., principally from the works of Simkin, et al, (1981), as well as documented direct eruption data from volcanic observatories about the world. *"ERUPTION"* performs analysis on loaded volcano eruption data from both historical and current eruption data. It then produces three forecasts; a statistically projected next eruption year, the next forecasted beginning eruption year with an $\geq 50\%$ probability of eruption occurrence, and finally, the next forecasted beginning eruption year with an $\geq 95\%$ probability of eruption occurrence. As for the case of the $\geq 50\%$ and $\geq 95\%$ probability calculations, *"ERUPTION"* solves this inequality for n years in the future where n is the minimum forecasted years for an eruption to occur beyond its last eruption with 50 and 95 percent probability respectively.

The Poisson distribution is a good model for describing pheonomena where the probability of occurrence is small and constant. It arises as the exact model underlying various physical phenomena such as is the case with volcanic eruptions which involve time. It is also an approximation where the number of trials, n, is large as is the case of volcanoes where hundreds and even thousands of years pass before an eruption. The probability of success (an eruption), p, is small. i.e., The Poisson distribution is an excellent distribution for rare events. It is a special case of the Binomial distribution. And, as De La Cruz-Reyna (1991) states in his work, *"If one concludes that well-sampled moderate-to-large magnitude sequences follow a Poisson distribution, then the basic features of Poissonian processes become fundamental in understanding the physics of volcanism. The analysis of published global data supports the notion that occurrence of eruptions can be accurately described as a simple Poisson process."*

In Poisson, one wishes to calculate the probability $P_r(x)$, of observing **r** events in the interval (0,x). Considered first is the probability of getting no events in (0,x), $P_0(x)$. Next considered is the probability $P_1(x)$, of getting one and only one event in the interval (0,x). One may therefore write

$$P_r = \frac{(\mu^r e^{-\mu})}{r!} \equiv p(r;\mu)$$

where $\mu = \lambda x$ = the expected value

and $\lambda$ = the probability density constant
x = the time span

The average eruption rate per year is given by

$$\mu = \frac{NOE}{(LEY - FEY)}$$

where, NOE = number of eruptions
LEY = last eruption year
FEY = first eruption year

One of the algorithms used by *"ERUPTION"* is a modified Poisson distribution equation. The Poisson distribution algorithm used by *"ERUPTION"* is given as

$$P_r = \frac{(\Delta\mu^r e^{-\Delta\mu})}{r!} \equiv p(r;\mu)$$

where $\Delta\mu$ = the change in eruption rate per year as given by

$$\Delta\mu = \frac{NOE}{(CY - LEY)}$$

where CY = the current year

The use of $\Delta\mu$ yields a more accurate assessment of the annual eruption rate since, as time continues, the rate also changes since the span of time has also changed.

With regard to the 50% and 95% profiles, *"ERUPTION"* calculates the number of years beyond the next statistical forecasted year in order to establish the year in which there is a 50% or 95% respectively chance of an event. If FE represents the next statistical forecasted year and N50 & N95 the number of years beyond FE then the 50% and 95% profiles may be calculated from

$$P_{(50\%)} = FE + N50 \quad \text{and} \quad P_{(95\%)} = FE + N95$$

where, $\quad$ N50 = (-1/$\mu$) * ln(.5)

and $\quad$ N95 = (-1/$\mu$) * ln(.05)

To date, the analysis for the years 1989 through 1994 and, to date through 1995, have been completed and are tabulated in Table 5–1. Results of the next statistical forecasted year prior to 1995 have been generally very favourable. Including the results of the $\geq$50% and $\geq$95% forecasts which also have shown remarkable accuracy. The most probable cause of the low accuracy percentiles for the next statistically forecasted eruption year prior to 1995 was discussed during a 1993 visit to volcano Piton de la Fournaise. An on-site discussion was held with Dr. Jean-Paul Toutain, then Director of the Observatoire Volcanologique du Piton de la Fournaise on Reunion Island, France, located in the Indian Ocean. The results of that discussion lead to the conclusion that there are, assuredly, probable "gaps" in the historical records, particularly as one traverses back in time and therefore contributes to this inaccuracy.

Table 5–2 presents the results of the 1994 forecast year and Table 5–3 presents the computer generated forecasts for the year 1995. The volcanoes listed in Table 5–3 are those volcanoes that *"ERUPTION"* has forecasted to erupt in 1995 based on the historical record, statistical analysis and pertinent recent input data. This does not mean to imply that other volcanoes will not erupt; they certainly shall. The initial testing of *"ERUPTION"* has been completed and the next objective of the programme is to now fully test the programme *"ERUPTION"* on its ability to forecast future volcanic eruptions in the categories of next forecasted year profile, the $\geq 50\%$ forecast profile and the $\geq 95\%$ forecast profile. An additional objective is to compare the actual eruption event results against the computer forecasted prediction results beginning with the year 1989. Subsequent forecasts will be accomplished for future years and the results shall be evaluated. Table 5–4 presents the 1996 forecast.

The experimental programme *"ERUPTION"* is still undergoing refinements. For example, in 1994 a Bayesian analysis, in conformance with a work completed by Dr. Chiang-Huang Ho of the University of Nevada, Las Vegas, in 1993, was incorporated into *"ERUPTION"*.

Revising probabilities when new information is obtained is an important part of probability analysis. Often, as is the case with most volcanoes assumed to be Poisson distributed, the initial or *prior* probability estimates are completed for a specific event of interest, i.e., the probability of an eruption for the current year. Then, some new additional information is obtained, a missed eruption, or the fact that another year transpires and there has been no eruptive event. Given this new information, the prior probabilities are updated by calculating the revised probabilities referred to as *posterior* probabilities. Bayes' Theorem provides a means for making such calculations and this theorem, along with the axioms suggested by the combining of Poisson distribution and negative binomial distribution and using a Bayesian analysis as they apply to volcanic eruptions (Ho, 1990), have been incorporated into *"ERUPTION"*.

When the Poisson process, as applied to volcanic eruptions, is expanded to accommodate a gamma mixing distribution on $\lambda$, there becomes an immediate consequence of this mixed Poisson model. The frequency distribution of distribution of eruptions in any given interval of equal time length follows a negative binomial distribution. The probability of x eruptions becomes:

$$P(x) = \frac{\Gamma(r+x)}{\Gamma(r)\,x!}\;[\alpha/(\alpha+1)]^r\;[1/(\alpha+1)]^x\;,\;x = 0, 1, 2, ......$$

where r and $\alpha$ are the shape and scale parameters of the gamma distribution respectively.

Treating the average eruption rate $\lambda$ as a random variable means that the probability distribution function $f(x,\lambda)$ is, in reality, a *conditional* probability. The condition being that $\lambda$ is in state $\lambda$. Therefore, when using a probability distribution for $\lambda$, it might be more suitable to use the notation $f(x|\lambda)$ for the data x. From the conditional distribution of x and the given (calculated) *prior* distribution for $\lambda$, the joint distribution of $(x,\lambda)$ can be calculated. Thus:

$$f(x,\lambda) \quad = \quad f(x|\lambda)g(\lambda)$$

where $g(\lambda)$ is the probability density function and the marginal or absolute distribution of x, with probability:

$$P(x) \quad = \quad E_g[f(x,\lambda)] \quad = \quad \int f(x|\lambda)g(\lambda)\,d\lambda$$

For the volcanoes being monitored by "*ERUPTION*", and assuming that $\lambda$ follows a gamma distribution, then

$$g(\lambda) \quad = \quad \frac{\underline{\alpha}^r\,\lambda^{r-1}\,e^{-\alpha\lambda}}{\Gamma(r)}\;;\; l > 0;\; r,\alpha > 0 \qquad (1)$$

where r and $\alpha$ are the shape and scale parameters respectively as previously mentioned, and

$$f(x|\lambda) = \frac{e^{-\lambda}\lambda^x}{x!}, \qquad x = 0, 1, ....$$

Therefore, from Equation (1) above, the absolute probability for the number of eruptions per unit of time interval is given by,

$$P(x) = \int_0^\infty \frac{e^{-\lambda}\lambda^x}{x!} \frac{\alpha^r}{\Gamma(r)} \lambda^{r-1} e^{-\alpha\lambda} d\lambda$$

$$= \frac{\Gamma(r+x)}{\Gamma(r) x!} [\alpha/(\alpha+1)]^r [1/(\alpha+1)]^x , \quad x = 0, 1, 2, ...$$

The mean and variance for the negative binomial distribution are given by:

$$E(x) = r/\alpha$$

$$\text{and} \qquad \text{Variance}(x) = r(\alpha+1)/\alpha^2 .$$

It has been at least one year now (1996) to see if the combined negative binomial and Poisson distributions along with the Bayesian analysis will have had any affect on the next statistical forecast accuracy of "*ERUPTION*". As can be observed in Table 5–1, the improvement factor has been a qualified successes. Another few years of testing will be necessary to verify these assumptions. Another improvement factor built into "*ERUPTION*" is the eruption event count. Although a particular volcano may erupt more than once during a given year, "*ERUPTION*" counts all the eruptions compiled for that volcano which has erupted at least once or more in a particular year.

In the interim, it is hoped that the programme will prove to be useful in the elusive ability to forecast long range pending volcanic eruptions, particularly of those volcanoes which, in their violence and sometimes destructive path, pose a serious threat to persons and property. A second objective of "*ERUPTION*" is intended to forecast those volcanoes which are forecasted to next erupt accurate to the nearest year. This is accomplished in order that, and in the hope that, surveillance, if not already being done on a particular volcano, may now be considered for that volcanic mountain, especially if a real and serious danger to the general populace at large exists.

Year 2006 Note:

This chapter is preserved in its original format in order to preserve the original prose written. *"ERUPTION"* has been updated extensively and is now called **Eruption Pro 10.6**. The details and methodologies used by **Eruption Pro 10.6** can be had by visiting our website at http://www.swvrc.org or e-mailing us at swvrc@usa.net.

## TABLE 5–1

## SUMMARY OF ANNUAL VOLCANO ERUPTION
## FORECASTS TO DATE AS GENERATED BY *"ERUPTION"*

| | CURRENT FORECASTS | | | $\geq 50\%$ FORECASTS | | | | $\geq 95\%$ FORECASTS | | | # OF |
|---|---|---|---|---|---|---|---|---|---|---|---|
| YEAR | FE | AE | PCT | 5FE | 5AE | PCT | 9FE | 9AE | PCT | | EVENTS |
| 1989 | 2 | 0 | 00.0 | 18 | 17 | 94.4 | 5 | 4 | 80.0 | | 46 |
| 1990 | 3 | 2 | 66.7 | 7 | 6 | 85.7 | 0 | 0 | 100.0 | | 32 |
| 1991 | 4 | 3 | 75.0 | 15 | 13 | 86.7 | 6 | 6 | 100.0 | | 48 |
| 1992 | 6 | 0 | 00.0 | 5 | 4 | 80.0 | 5 | 4 | 80.0 | | 55 |
| 1993 | 6 | 2 | 33.3 | 10 | 10 | 100.0 | 5 | 3 | 60.0 | | 48 |
| 1994 | 5 | 0 | 00.0 | 15 | 14 | 93.3 | 1 | 0 | 00.0 | | 61 |
| 1995[+] | 0 | 0 | 100.0 | 4 | 4 | 100.0 | 1 | 1 | 100.0 | | 41 |

Key:   FE = Forecasted Events              AE = Actual Events
PCT = Percent                         5FE = $\geq$50% Forecasted Events
5AE = $\geq$50% Actual Events      9FE = $\geq$95% Forecasted Events
9AE = $\geq$95% Actual Events

[+] As of 31 October 1995 per the Global Volcanism Network, Washington,
D.C., United States.

## TABLE 5–2a

## 1994 VOLCANO ERUPTION FORECAST RESULTS

VOLCANOES ERUPTING* IN 1994 WERE:

| VOLCANO NAME | LOCATION | LAT | LONG |
|---|---|---|---|
| +BEZYMIANNY | RUSSIA | 56.04N | 160.43E |
| +MERAPI | SUMATRA | 00.38S | 100.47E |
| +SEMERU | JAVA | 08.11S | 112.92E |
| KANAGA | ALEUTIAN IS. | 51.92N | 177.17W |
| MANAM | NEW GUINEA | 04.10S | 145.06E |
| LANGILA | NEW GUINEA | 05.53S | 148.42E |
| STROMBOLI | ITALY | 56.17N | 159.38W |
| SAKURA-JIMA | JAPAN | 31.58N | 130.67E |
| ARENAL | COSTA RICA | 10.47N | 085.73W |
| UNZEN | JAPAN | 32.75N | 130.30E |
| PACAYA | GUATEMALA | 14.38N | 090.60W |
| WHITE IS. | NEW ZEALAND | 37.52S | 177.18E |
| +RUAPEHU | NEW ZEALAND | 39.28S | 175.57E |
| MASAYA | NICARAGUA | 11.98N | 086.15W |
| COLIMA | MEXICO | 19.51N | 103.61W |
| +MOMOTOMBO | NICARAGUA | 12.42N | 086.53W |
| LASCAR | CHILE | 27.37S | 067.73W |
| NEVADO DEL RUIZ | COLUMBIA | 04.89N | 075.32W |
| LLAIMA | CHILE | 38.70S | 071.70W |
| CLEVELAND | ALEUTIAN IS. | 52.49N | 169.57W |
| MERAPI | INDONESIA | 07.45S | 110.44E |
| +SABANCAYA | PERU | 15.78S | 071.85W |
| POAS | COSTA RICA | 10.46N | 084.71W |
| ULAWUN | NEW GUINEA | 05.04S | 151.34E |
| KRAKATAU | INDONESIA | 16.10S | 105.42E |
| SHEVELUCH | RUSSIA | 56.38N | 161.09E |
| +NYAMURAGIRA | CENTRAL AFRICA | 01.38S | 029.20E |
| GAMALAMA | INDONESIA | 00.80N | 127.33E |
| +RINJANI | INDONESIA | 08.42S | 116.47E |
| TELICA | NICARAGUA | 12.60N | 086.84W |
| +BATUR | INDONESIA | 08.24S | 115.38E |
| +NYIRAGONGO | CENTRAL AFRICA | 01.52S | 029.25E |
| +TULUMAN | NEW GUINEA | 02.45S | 147.32E |

# TABLE 5–2a

# 1994 VOLCANO ERUPTION FORECAST RESULTS (Cont'd)

**VOLCANOES ERUPTING\* IN 1994 WERE:**

| VOLCANO NAME | LOCATION | LAT | LONG |
|---|---|---|---|
| ETNA | ITALY | 37.73N | 015.00E |
| KAWA IJEN | JAVA | 08.06S | 114.24E |
| KLIUCHEVSKOI | RUSSIA | 56.06N | 160.64E |
| ASO | JAPAN | 32.88N | 131.10E |
| +POPOCATEPETL | MEXICO | 19.02N | 098.62W |
| +RABAUL | NEW GUINEA | 04.27S | 152.20E |
| VILLARICA | CHILE | 39.42S | 071.95W |
| +BULUSAN | PHILIPPINES | 12.77N | 124.05E |
| DUKONO | INDONESIA | 01.70N | 127.87E |
| IRAZU | COSTA RICA | 09.98N | 083.85W |

Symbols Key:

\*   =   Forecasted to erupt in 1994.

+   =   Forecasted to erupt with $\geq$ 50% probability.

@   =   Forecasted to erupt with $\geq$ 95% probability.

## TABLE 5–2b

## 1994 VOLCANO ERUPTION FORECAST RESULTS

| | | | | |
|---|---|---|---|---|
| Volcanoes forecasted to erupt in 1994 | = | 6 | Actual = | 0 |
| Volcanoes forecasted to erupt with $\geq 50\%$ probability in 1994 | = | 15 | Actual = | 14 |
| Volcanoes forecasted to erupt with $\geq 95\%$ probability in 1994 | = | 1 | Actual = | 0 |

Percent accuracy of forecasted eruptions     =     0.000

Percent accuracy of $\geq 50\%$ forecasted eruptions =    93.333

Percent accuracy of $\geq 95\%$ forecasted eruptions =     0.000

Overall accuracy percentage      =     66.667

Number of volcanoes erupting in 1994    =     61

## TABLE 5–3a

## THE 1995 VOLCANO ERUPTION FORECAST

**VOLCANOES FORECASTED TO ERUPT IN 1995 ACCORDING TO**
***"ERUPTION"* ARE:**

| VOLCANO NAME | LOCATION | LAT | LONG |
|---|---|---|---|
| ***** NONE ***** | | | |

**VOLCANOES FORECASTED TO ERUPT IN 1995 WITH ≥ 50% PROBABILITY**
**ACCORDING TO *"ERUPTION"* ARE:**

| VOLCANO NAME | LOCATION | LAT | LONG |
|---|---|---|---|
| FOGO | CAPE VERDE IS. | 14.95N | 024.35W |
| METIS SHOAL | TONGA IS. | 19.18S | 174.87W |
| AOBA | SO. PACIFIC | 15.40S | 167.83E |
| KUJU | JAPAN | 33.08N | 131.25E |

**VOLCANOES FORECASTED TO ERUPT IN 1995 WITH ≥ 95% PROBABILITY**
**ACCORDING TO *"ERUPTION"* ARE:**

| VOLCANO NAME | LOCATION | LAT | LONG |
|---|---|---|---|
| TENGGER | JAVA | 07.94S | 112.95E |

## TABLE 5–3b

## 1995 VOLCANO ERUPTION FORECAST RESULTS

| | | | | |
|---|---|---|---|---|
| Volcanoes forecasted to erupt in 1995 | = | 0 | Actual = | 0 |
| Volcanoes forecasted to erupt with $\geq 50\%$ probability in 1995 | = | 4 | Actual = | 4 |
| Volcanoes forecasted to erupt with $\geq 95\%$ probability in 1995 | = | 1 | Actual = | 1 |

| | | |
|---|---|---|
| Percent accuracy of forecasted eruptions | = | 100.000 |
| Percent accuracy of $\geq 50\%$ forecasted eruptions | = | 100.000 |
| Percent accuracy of $\geq 95\%$ forecasted eruptions | = | 100.000 |
| Overall accuracy percentage | = | 100.000 |
| Number of volcanoes erupting in 1995 | = | 41 |

# TABLE 5–4

## THE 1996 VOLCANO ERUPTION FORECAST

**VOLCANOES FORECASTED TO ERUPT IN 1996 ACCORDING TO**
*"ERUPTION"* **ARE:**

| VOLCANO NAME | LOCATION | LAT | LONG |
|---|---|---|---|
| HIDSON | CHILE | 45.90S | 072.97W |
| KUWAE | SO. PACIFIC | 16.83S | 168.54E |
| LONQUIMAY | CHILE | 38.37S | 071.58W |
| SOUFRIERE HILLS | WEST INDIES | 16.72N | 062.18W |

**VOLCANOES FORECASTED TO ERUPT IN 1995 WITH ≥ 50% PROBABILITY**
**ACCORDING TO** *"ERUPTION"* **ARE:**

| VOLCANO NAME | LOCATION | LAT | LONG |
|---|---|---|---|
| AGUNG | INDONESIA | 08.34S | 115.51E |
| AKAN | JAPAN | 43.38N | 144.02E |
| FUEGO | GUATEMALA | 14.47N | 090.88W |
| ILIBOLENG | INDONESIA | 08.34S | 123.26E |
| MARAPI | SUMATRA | 00.38S | 100.47E |
| NEVADOS DE CHILLAN | CHILE | 36.86S | 071.38W |
| RITTER IS. | NEW GUINEA | 05.52S | 148.12E |

**VOLCANOES FORECASTED TO ERUPT IN 1995 WITH ≥ 95% PROBABILITY**
**ACCORDING TO** *"ERUPTION"* **ARE:**

| VOLCANO NAME | LOCATION | LAT | LONG |
|---|---|---|---|
| DEMPO | SUMATRA | 04.03S | 103.13E |
| NISHINO-SHIMA | JAPAN | 27.43N | 140.88E |

# GLOSSARY

**Aa**
A type of lava having a jagged, rough edge and surface.

**Active Volcano**
Generally volcanoes that have erupted in recorded history or are currently erupting.

**Aerosol**
A suspension of solid particulates or fine liquid in air.

**Andesite**
A gray coloured volcanic rock common to strato-volcanoes having a silica content between basalt and dacite.

**Ash**
In the volcanic sense, fine fragments of lava or rock, about dust size, formed by volcanic explosions.

**Ash Cloud**
A cloud of ash formed by a volcanic eruption or by a pyroclastic flow.

**Avalanche**
A large quantity of rocks, volcanic debris, earth, etc., descending swiftly down a mountain side.

**Basalt**
A type of lava, dark in colour, rich in iron and magnesium and containing approximately 50% silica.

**Caldera**
A large bowl or basin shaped depression at a volcano's summit generally formed by collapse.

**Cinder Cone**
A relatively steep mound or hill formed by the accumulation of cinders, usually around a vent of cinders and other volcanic fragments thrown out during an eruption.

**Composite Volcano**      Same as a Stratovolcano.

**Compressional Margin**      The edges of convergence of two tectonic plates.

**Conduit**      A tube or crack through which lava moves.

**Continental Crust**      The solid outer layer of the Earth beneath the continents. The continental crust is thicker and less dense than the oceanic crust. The continental crust is approximately 25 kilometres in thickness.

**Continental Drift**      The theory that the slow, relative movements of the continents is caused by horizontal movements of the Earth's surface.

**Crater**      A bowl-shaped depression around the mouth of a volcano.

**Curtain of Fire**      A wall of lava fountains that are erupting along a fissure.

**Dacite**      A light-coloured volcanic rock with intermediate silica composition ranging between rhyolite and andesite.

**Dike**      A mass of intrusive rock, usually blade-shaped, that cuts layers of surrounding country rock.

**Dormant Volcano**      A volcano that is not currently erupting but is likely to do so in the future.

**Earthquake Swarm**      A sequence of earthquakes, closely spaced in time, of approximately the same magnitude as opposed to a sequence of strong earthquakes with a series of diminishing aftershocks.

**Earthquake Wave**      A vibrational wave produced by an earthquake.

| | |
|---|---|
| **Effusive Eruption** | An eruption consisting mostly of lava flows as opposed to an explosive eruption. |
| **Eruption Cloud** | A cloud consisting of gases, ash, and other volcanic fragments which are generated by a volcanic eruption. |
| **Explosive Eruption** | A sudden expansion of gases ladened with volcanic fragments and other volcanic debris. |
| **Extensional Margin** | The edges of tectonic plates that are moving apart from one another. |
| **Extinct Volcano** | A volcano that has not erupted in historic times or is not expected to erupt again in the future. |
| **Fault** | A crack or fracture in the Earth's crust along which there has been evidence of movement. |
| **Fault Scarp** | A cliff formed along a fault caused by movement of the fault. |
| **Feldspar** | A light-coloured mineral composed principally of oxygen, aluminum, and silicon. |
| **Felsic Crust** | Crust that is made up of a light coloured igneous rock that is poor in iron and magnesium and contains abundant feldspars and quartz. |
| **Fissure** | A large blade-shaped fracture in the Earth's crust. |
| **Flank Eruption** | An eruption that vents from the side of a volcano instead of from the summit of the volcano. |
| **Fumarole** | An opening (vent) in the surrounding earth from which volcanic gases and steam are emitted. |
| **Geophysics** | The study of the physical and mechanical aspects of the science of geology. |

**Geothermal Energy**  The energy which is derived from the Earth's internal heat source.

**Geothermal Gradient**  The rate of change in temperature with depth in the Earth.

**Geothermal Power**  The power generated by the heat energy of the Earth.

**Granite**  A course-grained igneous rock composed chiefly of quartz and feldspar.

**Guyot**  A flat-topped submarine mountain or seamount.

**Harmonic Tremor**  A volcanic tremor that has a relatively steady state frequency and amplitude.

**Holocene**  The period of geologic time since the last major ice age; approximately 10,000 years ago to the present day.

**Hot-Spot Volcano**  A volcano related to a persistent heat source in the mantle.

**Hydrothermal**  An area that lies underground containing porous rock that contains a reservoir of hot water.

**Igneous Rock**  Magma or lava that has cooled and solidified either below or above the Earth's surface.

**Ignimbrite**  A widespread deposit that has been left by a large pyroclastic flow.

**Intrusion**  A body of rock formed by magma that has forced its way into the surrounding host rock and then cools. It is also the process of forming such a body of rock.

| | |
|---|---|
| **Island Arc** | A curving chain of volcanic islands, such as the Hawaiian chain, formed at compressional plate margins. |
| **Lahar** | A mudflow created by volcanic activity such as the melting of ice and snow. |
| **Lava** | Magma that has reached the surface of the Earth. It is also what the resulting rock is called after cooling. |
| **Lava Channel** | A swift-moving, incandescent portion of an active lava flow or its solidified remains. |
| **Lava Dome** | A mass of viscous lava, steep-sided, usually with a rounded shaped top which covers a volcanic vent. |
| **Lava Flow** | A stream of molten rock, effusive in nature that moves down the side of a volcano. |
| **Lava Fountain** | A very rapid jet of incandescent lava sprayed from a volcanic vent caused by the rapid expansion of gases. |
| **Lava Lake** | Literally, a lake of molten lava within a volcanic crater or depression. |
| **Lava Tube** | A tunnel beneath the surface of a solidified lava flow. Also the cave formed by the emptying of a tunnel as the eruption ceases or shifts direction. |
| **Magma** | Molten rock within the Earth. Magma that reaches the surface of the Earth is refereed to as lava. |
| **Magma Chamber** | The underground reservoir in which magma is stored. |

**Magmatic Gases**      Various gases such as carbon dioxide, hydrogen sulfide and water that is contained (dissolved) in magma.

**Mantle**      The zone of the Earth below the crust to a depth of approximately 3500 kilometres. Also that area that is located above the core of the Earth.

**Microearthquake**      An earthquake that is too small to be felt at the surface of the Earth but is detectable with a seismograph.

**Mudflow**      A water-saturated mixture of mud and debris that flows downslope under the force of gravity. Also known as a Lahar.

**Neck**      A vertical intrusion, usually seen as an erosional remnant, that depicts a former volcanic conduit.

**Normal Fault**      An inclined fault whereby the upper block moves downward relative to the lower block.

**Nuée Ardente**      (Fr) A dense "glowing cloud" of volcanic gas and ash that erupts from a volcano and moves swiftly down the slopes as a pyroclastic flow.81

**Obsidian**      A dark coloured or black volcanic glass that is generally composed of rhyolite.

**Oceanic Crust**      The crust of the Earth where it lies under the oceans and without the layer of granite that forms the continents. It is generally approximately 5 kilometres in thickness.

**Oceanic Ridge**      A major mountain range that lies completely underwater.

**Olivine**      An olive-green mineral composed of silicon, iron, magnesium and oxygen.

**Ore**

Any rock material from which minerals of commercial value can be extracted.

**Pahoehoe**

A basaltic lava with a smooth "ropy" or billowy textured surface.

**Partial**

A stage of cooling magma where it is composed of partly solid crystallization crystals and partly liquid rock.

**Partial Melt**

A stage of melting rock when it is partly solid and partly liquid crystals.

**Pelee's Hair**

Natural spun glass in strands formed generally in lava fountains.

**Pillow Lava**

Rounded, bag or sac like bodies of lava that form underwater as it is extruded.

**Plate Tectonics**

The theory that the Earth's crust is broken into many pieces, about a dozen or so plates, that slowly move about in relation to one another.

**Plume**

A column of hot, plastic rock rising from well within the mantle to form what are known as "hot-spot" volcanoes.

**Pluton**

A large igneous intrusion that cools and solidifies beneath the surface of the Earth.

**Pumice**

A form of volcanic glass filled with so much gas bubbles and holes that it resembles a sponge and is very light in weight.

**Pyroclastic Deposit**

The deposit of volcanic fragments from a pyroclastic flow.

**Pyroclastic Flow**

Same as a nuée ardente.

**Quartz**              A rock forming mineral composed chiefly of silicon and oxygen.

**Repose Time**         The time interval between eruptions.

**Rift System**         The oceanic ridges, greater than 60,000 kilometres in length, where the tectonic plates are separating and new crust is being formed. An example of the surface counterpart is the East African Ridge.

**Rift Volcano**        A volcano along a rift system.

**Rift Zone**           A region whereby the crust is separating and pulling apart.

**Ring of Fire**        The region of converging tectonic plate margins, with the resulting earthquakes and volcanoes, that surrounds the Pacific Ocean.

**Rhyolite**            A fine-grained volcanic rock, similar in nature to granite, but with a high silica composition.

**Seafloor Spreading**  The creation of new seafloor at the oceanic ridges as the tectonic plates separate from one another.

**Seamount**            A mountain, usually alone and volcanic, that lies underwater.

**Seismic Wave**        Same as an earthquake wave.

**Seismograph**         An instrument used for recording seismic waves in the crust of the Earth.

**Seismology**          The study of seismic waves, earthquakes, and the interior of the Earth's structure.

**Shield Volcano**     A volcano which is built by successive flows of fluid basaltic lava and forms a dome shaped structure with gentle sloping sides.

**Silica**     A chemical composition comprised of silicon and oxygen.

**Silicate Mineral**     A mineral composed mainly of silicon and oxygen.

**Silcic**     A term for volcanic rock or magma that is rich in silica and is similar to rhyolite.

**Solfatara**     A fumarole that has gases that are principally sulfurous.

**Spatter Cone**     A cone built up around a vent by fragments of still molten lava that weld into a large mass.

**Stratovolcano**     A steep sided volcanic cone built by a combination of lava flows and pyroclastic flows from explosive eruptions.

**Strike-Slip Fault**     A predominantly vertical fault with sideslipping, horizontal displacement.

**Subduction-Type**     A volcano that occurs just inland from a subduction zone such as those volcanoes along the Cascade range.

**Subduction Zone**     The zone where two tectonic plates converge generally with one plate overriding the other plate.

**Surge**     A short timed increase in the velocity and volume of a lava flow.

**Talus**     A gathering of rock debris at the base of a cliff or steep slope.

**Tephra**   Material of all sizes and types that erupts from a volcano and is usually deposited by airfall. Also another name for pyroclastic deposits.

**Thermal Gradient**   The rate of change of temperature with increase in depth or distance.

**Thrust Fault**   A gently inclined fault whose upper side moves relatively upward.

**Tidal Wave**   Same as a tsunami.

**Transform Fault**   A strike-slip fault that connects the offsets of a mid-oceanic ridge.

**Tsunami**   A giant sea wave produced by an underwater earthquake, landslide or volcanic eruption.

**Vein**   An opening in the surface of the Earth through which volcanic material are erupted.

**Viscosity**   A measure of the resistance to flow in a liquid.

**Volcanic Block**   Volcanic debris fragments, generally larger than 64mm in size, that are thrown out during an explosive eruption.

**Volcanic Bomb**   A lump of molten lava which is thrown out of a volcano and takes on a rounded shape.

**Volcanic Cinder**   A lava fragment approximately 1 cm in diameter.

**Volcanic Complex**   A persistent volcanic event area that has, over a period of time, built a complicated mixture of volcanic landforms.

**Volcanic Dust**   Fine particulates of volcanic ash.

**Volcanic Front**              A line of volcanoes that are closest to an oceanic trench.

**Volcanic Tremor**             A continuous vibration of the ground, detectable by seismograph that is associated with volcanic eruptions and other subsurface volcanic activity.

**Welded Tuff**                 Pyroclastic deposits, so hot when formed, that the fragments weld together forming a solid rock.

# REFERENCES

1. Detrick, R.S., Von Herzen, R.R., Crough, S.T., Epp, D. and Fehn, U., *"Heat flow on the Hawaiian swell and lithospheric reheating."*, Nature, 292, 142–143, 1981.

2. Simkin,L., McClelland, L., Bridge, D., Newhall. C., Latter, J.H., *"Volcanoes of the World"*, Washington, D.C. : Smithsonian Institution, 1981.

3. Simkin, T., Siebert, L., McClleland, L., May, *"Volcanoes of the World"*, 1984 Supplement, Washington, D.C.: Smithsonian Institution, 1984.

4. Watts, A.B., Bodine, J.H., Steckler, M.S., *"Observations of flexure of the lithosphere at seamounts."*, J. Geophysics.Res., 89, 13, 11152–11170, 1984.

5. Wickman, F. E., 1966, *"Repose patterns of volcanoes". I. Volcanic eruptions regarded as random phenomena."*, Ark:Mineral. Geol.,4:291–301.

6. De La Cruz-Reyna, S., 1991. *"Poisson-distributed patterns of explosive eruptive activity."*, Bulletin of Volcanology., 54:57–67.

7. Observatoire Volcanologique du Piton de la Fournasie, Réunion Island, Indian Ocean, France, Eruption data.

8. Ho, Chih-Ssiang, 1990, *"Bayesian analysis of volcanic eruptions"*, Journal of Volcanology and Geothermal Research, 43: 91–98

9. Dubois, J. & Cheminée, J. L., 1991, *"Are Sequences of Volcanic Eruptions Deterministically Chaotic?"*, Journal of Geophysical Research, Vol. 96, No.B7, Pgs. 11,931–11,945

10. Trombley, R. B., 1990, *"Computer modeling of statistical explosive patterns and the probability of volcanic event forecasting."*, Digital Equipment Corporation, U.S. Education Services, white paper.

11. Sempéré, Jean Christophe & Klein, Emily, M., *"New Insights in Crustal Accretion Expected From Indian Ocean Centers"*, EOS, Transactions, American Geophysical Union, Vol. 76, No.11, March 14, 1995, Pgs. 113 & 116.

12. Bérnard, Roland & Kraft, Maurice, *"Au Coeur de la Fournaise"*, Éditions Nouraualt/Bénard, Pgs. 44–51.

13. Stieltjes, L., *"Historical volcanic activity of Piton de la Fournaise (Réunion Island, western Indian Ocean)—Statistical data and geological implication."*, 27e Congrès Géologique International, Moscou, August, 1984.

14. Hamilton, W.L., *"Tidal cycles of volcanic eruptions: Fortnightly to 19 yearly periods"*, J. Geophysics, Res., 78(17):3363–3375, 1973

15. Martin, D.P. & Rose, W.L., *"Behavioural patterns of Fuego volcano, Guatemala"*, J. Volcanol.Geotherm.Res., 10:67–81, 1981

16. Mauk, F.J. & Johnston, M.J.S., *"On the triggering of volcanic eruptions by earth tides"*, J. Geophysics.Res., 78:3356–3362

17. Toutain, Jean-Paul, *"The 1992 Eruption of Piton de la Fournaise, Reunion Island, France"*, World Organizations of Volcanic Obervatories, Quarterly Newsletter, Number 2, Spring 1993, Pgs. 14–17

18. Trombley, R.B., *"A Computer Based Long-Range Volcano Eruption Forecasting Programme, "ERUPTION""*, EOS, Transactions, American Geophysical Union (AGU), Vol 76, No. 46, Supplement, 7 November 1995

19. Simkin, L., Siebert, L., *"Volcanoes of the World"*, Washington, D.C. : 2nd Edition, Smithsonian Institution, 1993.

# ABOUT THE AUTHOR

Dr. Robert B. Trombley is an award winning scientist and is currently an Associate Professor of Mathematics & Computer Sciences, at the DeVRY Institute of Technology in Phoenix, Arizona. In 1993, he was the visiting volcanologist at the Observatoire Volcanologique du Piton de la Fournaise on France's Réunion Island located in the Indian Ocean completing investigative research on volcano Piton de la Fournaise. He has been an educator and scientist for over 30 years and is both a member and contributor to the American Geophysical Union (AGU). He is the winner of the coveted Bauch-Lomb Honorary Science Medal awarded for outstanding contributions in the field of science. In 1988, Dr. Trombley was elected to *"Who's Who In America"*— Midwest Edition, primarily for his contributions to aerospace and defence projects.

Dr. Trombley received his undergraduate degree in electronic engineering from the Lawrence Technological University in 1965 and also received an LL.B from LaSalle University in 1973. He obtained his Ph.D. in 1974 from University of Dallas and completed additional post-doctorate training in particle physics fro the University of California Los Angeles. Dr. Trombley has also completed post-doctorate training in geology and paleogeology from the University of New Mexico.

Dr. Trombley has been actively pursuing the science of volcanology for the past twelve years, concentrating primarily on the very difficult and controversial task of volcanic eruption forecasting techniques. He has recently developed a new computer software package where he has loaded post-Holocene historic and current eruption data on 448 active volcanoes throughout the world and has performed a statistical analysis of that data. To date, there have been very favourable and accurate forecasts made as a result, particularly in the 50 and 95 percent probability of eruption calculation. Research into volcanic eruption forecasting continues to occupy a considerable amount of scientific research and software development.

# APPENDIX A
# PHOTOGRAPHIC PLATES

*Plate 1*      *The Grand Bulé*
*(Photographed by Jean-Paul Toutain.)*

*Plate 2*       *Typical Red Ibiscus Flowers Of Réunion Island*
              *(Photographed by R.B. Trombley)*

*Plate 3*        *Typical Residential Housing About Réunion Island*
                 *(Photographed by R.B. Trombley)*

*Plate 4*       *Author & The Caldera Of Piton des Neiges*
              *(Photographed by Jean-PaulToutain)*

*Plate 5*        *The 1986 Eruption Within Crater Dolomieu*
                 *(Photographed by Roland Bernard)*

*Plate 6*        *Crater Dolomieu In Eruption At Sunrise*
                 *(Photographed by Roland Bernard)*

*Plate 7*        *A Vent Eruption Of Pahoehoe Type Lava*
              *(Photographed by R.B. Trombley)*

Plate 8     Crater Dolomieu in Erupton Clearly Showing Aa
            Lava In Centre Of View And Pahoehoe Type Lava In
            The Foreground (Photographed by Roland Bernard)

*Plate 9*  *Spectaclular Vent Eruption Of Crater Dolomieu*
*(Photographed by Roland Bernard)*

*Plate 10      Vent Eruption Of Piton de la Fournaise*
*(Photographed by R.B. Trombley)*

*Plate 11*  *Spectacular Night View Of Crater Dolomieu In Eruption Showing Braided Lava Flowing DownTo The Indian Ocean (Photographed by Roland Bernard)*

*Plate 12*     *Another View Of Piton de la Fournaise In Eruption At Sunset (Photographed by Roland Bernard)*

*Plate 13*  *A Carpet Of Aa Type Lava Flowing Down The Grand Brulé (Photographed by R.B. Trombley)*

*Plate 14*        *Close-up View Of Molten Pahoehoe Type Lava*
*(Photographed by Roland Bernard)*

*Plate 15      A Fumarolic Vent Showing Sulphur Deposits*
*AboutThe Opening*
*(Photographed by R.B. Trombley)*

*Plate 16*     *Directly Inside Crater Dolomieu Within The Enclos*
          *(Photographed by R. B. Trombley)*

*Plate 17*     *Another View From Within Crater Dolomieu Clearly*
*Showing Venting*
*(Photographed by R.B. Trombley)*

*Plate 18*       *Final View Of Crater Dolomieu Showing The Total Desolation And Destruction (Photographed by R.B. Trombley)*

Plate 19          *Sign Marking The Entrance To The Volcano*
                  *Observatory (Photographed by R.B. Trombley)*

*Plate 20*     *The Observatory Proper Showing The Major*
*Buildings With Extinct Volcano Piton des Neiges In*
*The Background (Photographed by R.B. Trombley)*

*Plate 21*    *The Visiting Scientists Quarters At L'Observatoire*
*Volcanologique du Piton de la Fournaise*
*(Photographed by R.B. Trombley)*

*Plate 22*     *The Meteorological Station At The Observatory.*
*Fencing Is To Keep Animals (Cows) Away From The*
*Equipment (Photographed by R.B. Trombley)*

*Plate 23*     *The Antennae Which Receives Data From The Top Of*
*Piton de la Fournaise*
*(Photographed by R.B. Trombley)*

*Plate 24*    *The ARGOS Satellite System Antenna*
*(Photographed by R.B. Trombley)*

*Plate 25*     *A View Of The Data Recording And Seismic*
*Monitoring Equipment Showing The Tape Recorders*
*And Strip Recording Seismograph*
*(Photographed by R.B. Trombley)*

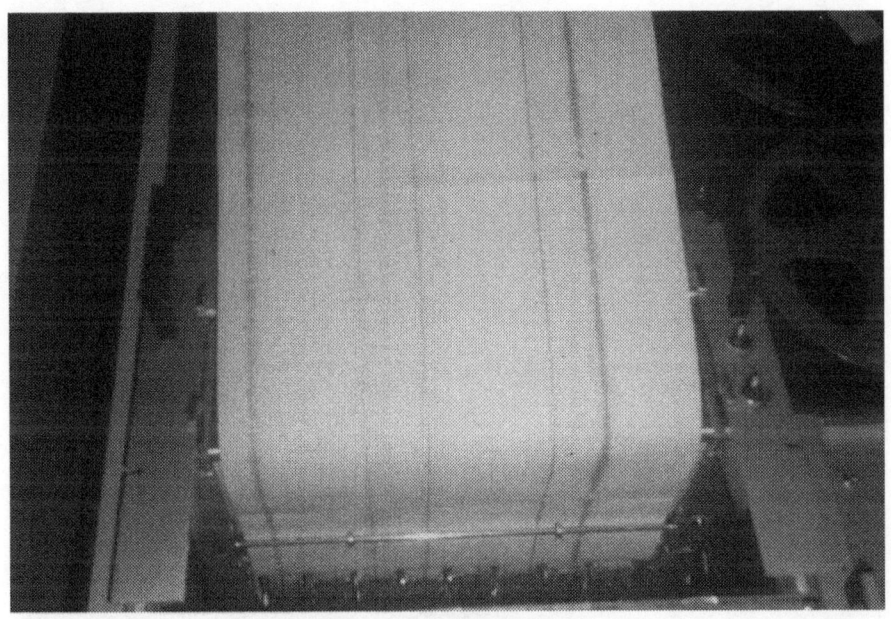

*Plate 26*    *The Strip Recording Seismograph*
*(Photographed by R.B. Trombley)*

*Plate 27*      *Transportation To The Top Of Piton de la*
                *Fournaise Is Provided By A Local Helicopter Service*
                *(Photographed by R.B. Trombley)*

*Plate 28*     *The Seismic Equipment, Radio Transmitting Antennae
And Solar Panels Just Outside Crater Dolimieu Atop
Piton de la Fournaise (Photographed by R.B.
Trombley)*

# APPENDIX B
# VOLCANIC EXPLOSIVE INDEX (VEI) CHART

| VEI | Description | Volume of Ejecta($m^3$) | Column Height (Km) | Qualitative Description | Classification |
|-----|-------------|-------------------------|--------------------|--------------------------|----------------|
| 0 | Non-Explosive | $<10^4$ | $<0.1$ | Gentle Effusive | Hawaiian |
| 1 | Small | $10^4$-$10^6$ | 0.1 - 1 | Gentle Effusive | Hawaiian-Strombolian |
| 2 | Moderate | $10^6$-$10^7$ | 1 - 5 | Explosive | Strombolian |
| 3 | Mod-Large | $10^7$-$10^8$ | 3 - 15 | Explosive | Strombol-Vulcanian |
| 4 | Large | $10^8$-$10^9$ | 10 - 25 | Cataclysmic | Plinian-Vulcanian |
| 5 | Very Large | $10^9$-$10^{10}$ | $>25$ | Cataclysmic | Vulcanian-Plinian |
| 6 | Very Large | $10^{10}$-$10^{11}$ | $>25$ | Paroxysmal | Plinian-Ultra-Plinian |
| 7 | Very Large | $10^{11}$-$10^{12}$ | $>25$ | Colossal | Plinian-Ultra-Plinian |
| 8 | Very Large | $>10^{12}$ | $>25$ | Colossal | Ultra-Plinian |

# APPENDIX C

# ERUPTION HISTORY[1] OF PITON de la FOURNAISE

## Réunion Island, France—Indian Ocean
## 21.229S     055.713E

| Start Year | | Stop Year | | CERF | SIGC | ENPS | FLDS | FDMT | VEI |
|---|---|---|---|---|---|---|---|---|---|
| 1640 | ---- | 1649 | ---- | ---- | ---- | X--- | X--- | ---- | 2 |
| 1669 | ---- | 1672 | ---- | ---- | ---- | X--- | X--- | ---- | 2 |
| 1703 | ---- | 1705 | ---- | ---- | ---- | ---- | X---- | ---- | 0 |
| 1708 | ---- | ---- | ---- | --X- | ---- | ---- | X--- | ---- | 0 |
| 1709 | ---- | ---- | ---- | ---- | ---- | X--- | X--- | ---- | 2 |
| 1721 | 06-- | ---- | ---- | ---- | ---- | X--- | ---- | ---- | 2 |
| 1733 | ---- | ---- | ---- | ---- | ---- | ---- | X--- | ---- | 0 |
| 1734 | 0101 | 1734 | 0306 | ---- | ---- | ---- | X--- | ---- | 2 |
| 1734 | 12-- | ---- | ---- | ---- | ---- | ---- | X--- | ---- | 2 |
| 1751 | 06-- | ---- | ---- | ---- | ---- | X--- | X--- | ---- | 2 |
| 1753 | ---- | ---- | ---- | ---- | ---- | X--- | X--- | ---- | 2 |
| 1760 | 1215 | 1760 | 1229 | ---- | ---- | X--- | ---- | ---- | 2 |
| 1766 | 0514 | 1766 | 0526 | X--- | ---- | X--- | X--- | ---- | 2 |
| 1768 | ---- | ---- | ---- | -X-- | ---- | X--- | X--- | ---- | 2 |
| 1772 | 1118 | ---- | ---- | X--- | ---- | X--- | X--- | ---- | 2 |
| 1774 | ---- | ---- | ---- | -X-- | ---- | ---- | X--- | ---- | 0 |
| 1775 | ---- | ---- | ---- | ---- | ---- | X--- | X--- | ---- | 2 |
| 1776 | ---- | ---- | ---- | --X- | ---- | ---- | X--- | ---- | 0 |
| 1784 | ---- | 1785 | ---- | X--- | ---- | X--- | ---- | ---- | 2 |
| 1786 | 0605 | 1800 | ---- | ---- | ---- | X--- | X--- | ---- | 2 |
| 1787 | 0617 | 1787 | 0801 | XX-- | ---- | X--- | X--- | ---- | 2 |
| 1791 | 0626 | 1791 | 0727 | X--- | ---- | X--- | X-X- | ---- | 2 |
| 1800 | 1102 | 1800 | 1106 | --X- | ---- | ---- | X--- | ---- | 0 |
| 1801 | 10-- | 1802 | 12-- | X--- | ---- | X--- | X--- | ---- | 2 |

| Start Year | | Stop Year | | Eruption Characteristics | | | | | |
|---|---|---|---|---|---|---|---|---|---|
| | | | | CERF | SIGC | ENPS | FLDS | FDMT | VEI |
| 1807 | 0323 | 1807 | 0613 | xx-- | ---- | x--- | x--- | ---- | 2 |
| 1809 | 0717 | 1809 | 0808 | x--- | ---- | x--- | ---- | ---- | 2 |
| 1810 | 1120 | 1810 | 1128 | x--- | ---- | x--- | ---- | ---- | 2 |
| 1812 | 0805 | 1812 | 0925 | xx-- | ---- | x--- | x--- | ---- | 2 |
| 1813 | 0626 | 1813 | 1126 | x--- | ---- | x--- | x--- | ---- | 2 |
| 1814 | 0910 | 1814 | 1013 | x--- | ---- | x--- | x--- | ---- | 2 |
| 1815 | 0121 | 1815 | 0127 | x--- | ---- | x--- | ---- | ---- | 2 |
| 1815 | 0815 | 1815 | 0616 | xx-- | ---- | x--- | x--- | ---- | 2 |
| 1816 | 1215 | ---- | ---- | ---- | ---- | ---- | x--- | ---- | 0 |
| 1817 | 01-- | 1817 | 04-- | x--- | ---- | ---- | ---- | ---- | 0 |
| 1821 | 0227 | 1821 | 0409 | x--- | ---- | x--- | x--- | ---- | 2 |
| 1824 | ---- | ---- | ---- | ---- | ---- | ---- | x--- | ---- | 0 |
| 1830 | 10-- | ---- | ---- | ---- | ---- | ---- | x--- | ---- | 0 |
| 1832 | ---- | ---- | ---- | ---- | ---- | ---- | x--- | ---- | 0 |
| 1842 | 04-- | ---- | ---- | ---- | ---- | ---- | x--- | ---- | 0 |
| 1843 | ---- | ---- | ---- | ---- | ---- | ---- | x--- | ---- | 0 |
| 1844 | 0319 | 1844 | 0511 | x-x- | ---- | x--- | x--- | ---- | 2 |
| 1845 | ---- | ---- | ---- | ---- | ---- | ---- | x--- | ---- | 0 |
| 1846 | ---- | ---- | ---- | ---- | ---- | ---- | x--- | ---- | 0 |
| 1847 | ---- | ---- | ---- | ---- | ---- | ---- | x--- | ---- | 0 |
| 1848 | ---- | ---- | ---- | ---- | ---- | ---- | x--- | ---- | 0 |
| 1849 | ---- | ---- | ---- | ---- | ---- | ---- | x--- | ---- | 0 |
| 1850 | 10-- | ---- | ---- | ---- | ---- | ---- | x--- | ---- | 0 |
| 1858 | 1103 | 1858 | 1214 | x--- | ---- | x--- | x--- | ---- | 2 |
| 1859 | 0508 | 1859 | 0523 | x--- | ---- | x--- | x--- | ---- | 2 |
| 1860 | 0122 | 1860 | 0319 | x-x- | ---- | x--- | x--- | ---- | 2 |
| 1863 | 1226 | 1864 | 0129 | x--- | ---- | ---- | x--- | ---- | 0 |
| 1865 | 0205 | 1865 | 0210 | x--- | ---- | x--- | x--- | ---- | 2 |
| 1868 | ---- | ---- | ---- | ---- | ---- | ---- | x--- | ---- | 0 |
| 1870 | ---- | ---- | ---- | ---- | ---- | ---- | x--- | ---- | 0 |
| 1871 | 0621 | 1871 | 0705 | x--- | ---- | ---- | x--- | ---- | 0 |
| 1872 | ---- | ---- | ---- | ---- | ---- | ---- | x--- | ---- | 0 |
| 1874 | 0629 | 1874 | 1107 | x--- | ---- | x--- | x--- | ---- | 2 |
| 1875 | ---- | ---- | ---- | ---- | ---- | ---- | x--- | ---- | 0 |
| 1878 | 0314 | 1878 | 0330 | -x-- | ---- | ---- | x--- | ---- | 0 |
| 1882 | ---- | ---- | ---- | ---- | ---- | ---- | x--- | ---- | 0 |
| 1884 | 0204 | 1884 | 0205 | ---- | ---- | ---- | x--- | ---- | 0 |

| Start Year | | Stop Year | | Eruption Characteristics | | | | | |
|---|---|---|---|---|---|---|---|---|---|
| | | | | CERF | SIGC | ENPS | FLDS | FDMT | VEI |
| 1889 | 06-- | 1889 | 0811 | xx-- | ---- | x--- | x--- | ---- | 2 |
| 1890 | 0605 | 1891 | 0204 | -x-- | ---- | x--- | x--- | ---- | 2 |
| 1897 | 0124 | ---- | ---- | --x- | ---- | ---- | x--- | ---- | 0 |
| 1898 | 0114 | 1898 | 0120 | ---- | ---- | ---- | x--- | ---- | 2 |
| 1898 | 1126 | ---- | ---- | x--- | ---- | x--- | x--- | ---- | 2 |
| 1899 | 0213 | 1899 | 0708 | -xx- | ---- | x--- | x--- | ---- | 2 |
| 1900 | 0511 | 1900 | 0530 | -x-- | ---- | ---- | x--- | ---- | 0 |
| 1901 | 0221 | 1901 | 0225 | -x-- | ---- | ---- | x--- | ---- | 2 |
| 1901 | 0704 | 1901 | 0706 | -x-- | ---- | x--- | x--- | ---- | 2 |
| 1902 | 0813 | 1902 | 0816 | --x- | ---- | x--- | x--- | ---- | 2 |
| 1903 | ---- | ---- | ---- | ---- | ---- | ---- | x--- | ---- | 0 |
| 1904 | 0819 | 1904 | 1017 | -x-- | ---- | x--- | x--- | ---- | 2 |
| 1905 | 0215 | 1905 | 0216 | x--- | ---- | x--- | x--- | ---- | 2 |
| 1907 | 1129 | 1907 | 1205 | -x-- | ---- | ---- | x--- | ---- | 0 |
| 1908 | ---- | ---- | ---- | ---- | ---- | ---- | x--- | ---- | 0 |
| 1909 | 04-- | ---- | ---- | -x-- | ---- | x--- | x--- | ---- | 2 |
| 1910 | 1116 | 1910 | 1212 | -x-- | ---- | x--- | x--- | ---- | 2 |
| 1913 | 0710 | 1913 | 0803 | --x- | ---- | x--- | x--- | ---- | 2 |
| 1915 | 0722 | 1915 | 1118 | xx-- | ---- | x--- | x--- | ---- | 2 |
| 1917 | ---- | ---- | ---- | ---- | ---- | ---- | x--- | ---- | 0 |
| 1920 | 0628 | 1920 | 1018 | --x- | ---- | ---- | x--- | ---- | 0 |
| 1921 | 1127 | 1921 | 1203 | x--- | ---- | x--- | x--- | ---- | 2 |
| 1924 | 0903 | 1924 | 0912 | x--- | ---- | x--- | x--- | ---- | 2 |
| 1925 | 1230 | 1926 | 0420 | x-x- | ---- | x--- | x--- | ---- | 2 |
| 1926 | 0918 | 1927 | 0615 | --x- | ---- | ---- | x--- | ---- | 2 |
| 1929 | 1223 | 1929 | 1230 | x--- | ---- | x--- | x--- | ---- | 2 |
| 1930 | 0523 | 1930 | 0524 | x--- | ---- | ---- | x--- | ---- | 0 |
| 1931 | 0122 | 1931 | 0819 | xxx- | ---- | x--- | x--- | ---- | 2 |
| 1933 | 0607 | 1933 | 0615 | x--- | ---- | x--- | x--- | ---- | 2 |
| 1933 | 1111 | 1934 | 0401 | xx-- | ---- | x--- | x--- | ---- | 2 |
| 1937 | 0813 | 1937 | 0912 | --x- | ---- | x--- | x--- | ---- | 2 |
| 1938 | ---- | ---- | ---- | -x-- | ---- | x--- | x--- | ---- | 2 |
| 1938 | 1207 | 1939 | 0123 | -xx- | ---- | x--- | FLDS | ---- | 2 |
| 1942 | 10-- | --- | --- | --x- | ---- | x--- | x--- | ---- | 2 |
| 1943 | 0405 | 1943 | 0526 | -xx- | ---- | x--- | x--- | ---- | 2 |
| 1944 | 0416 | ---- | ---- | -x-- | ---- | x--- | x--- | ---- | 2 |
| 1945 | 0415 | 1945 | 0506 | -x-- | ---- | x--- | x--- | ---- | 2 |

| Start Year | | Stop Year | | Eruption Characteristics | | | | | |
|---|---|---|---|---|---|---|---|---|---|
| | | | | CERF | SIGC | ENPS | FLDS | FDMT | VEI |
| 1946 | 0618 | 1946 | 0705 | -x-- | ---- | x--- | x--- | ---- | 2 |
| 1948 | 0215 | 1948 | 0305 | -x-- | ---- | x--- | x--- | ---- | 2 |
| 1949 | 10-- | ---- | ---- | -x-- | ---- | x--- | x--- | ---- | 2 |
| 1950 | 0308 | 1950 | 0415 | -x-- | ---- | x--- | x--- | ---- | 2 |
| 1950 | 08-- | 1950 | 09-- | -x-- | ---- | x--- | x--- | ---- | 2 |
| 1952 | 0530 | 1952 | 0720 | -x-- | ---- | x--- | x--- | ---- | 2 |
| 1953 | 0313 | 1953 | 0708 | -xx- | ---- | x--- | x--- | ---- | 2 |
| 1955 | 0706 | 1957 | 0316 | xxx- | ---- | x--- | x-x- | ---- | 2 |
| 1957 | 0902 | 1957 | 1120 | -x-- | ---- | x--- | x--- | ---- | 2 |
| 1958 | 0530 | 1958 | 0920 | -x-- | ---- | x--- | x--- | ---- | 2 |
| 1959 | 0311 | 1959 | 0806 | -x-- | ---- | x--- | x--- | ---- | 2 |
| 1960 | 0111 | 1960 | 0310 | xxx- | ---- | x--- | x--- | ---- | 2 |
| 1961 | 04-- | ---- | ---- | -x-- | ---- | ---- | x--- | ---- | 0 |
| 1963 | 11-- | ---- | ---- | x--- | ---- | x--- | ---- | ---- | 2 |
| 1964 | 0501 | 1964 | 0507 | x--- | ---- | x--- | x--- | ---- | 2 |
| 1964 | 1221 | 1965 | 0215 | -xx- | ---- | x--- | x--- | ---- | 2 |
| 1966 | 03-- | 1966 | 05-- | --x- | ---- | ---- | x--- | ---- | 2 |
| 1972 | 0609 | 1972 | 1210 | -xx- | ---- | x--- | x--- | ---- | 2 |
| 1973 | 0510 | 1973 | 0528 | x--- | ---- | x--- | x--- | ---- | 2 |
| 1975 | 1104 | 1976 | 0406 | -xx- | ---- | x--- | xx-- | ---- | 1 |
| 1976 | 1102 | 1976 | 1103 | --x- | ---- | x--- | x--- | ---- | 1 |
| 1977 | 0324 | 1977 | 0416 | --x- | ---- | x--- | x--- | -x-- | 0 |
| 1977 | 1024 | 1977 | 1117 | --x- | ---- | x--- | x--- | ---- | 1 |
| 1979 | 0528 | 1979 | 0529 | --x- | ---- | ---- | x--- | ---- | 1 |
| 1979 | 0713 | 1979 | 0714 | --x- | ---- | ---- | x--- | ---- | 0 |
| 1981 | 0203 | 1981 | 0225 | --x- | ---- | ---- | x--- | ---- | 2 |
| 1981 | 0226 | 1981 | 0330 | --x- | ---- | ---- | x--- | ---- | 2 |
| 1981 | 0401 | 1981 | 0505 | --x- | ---- | ---- | x--- | ---- | 2 |
| 1983 | 1204 | 1984 | 0118 | --x- | ---- | x--- | x--- | ---- | 1 |
| 1984 | 0118 | 1984 | 0218 | --x- | ---- | ---- | x--- | ---- | 2 |
| 1985 | 0614 | 1985 | 0615 | --x- | ---- | ---- | x--- | ---- | 1 |
| 1985 | 0805 | 1985 | 0901 | --x- | ---- | ---- | x--- | ---- | 2 |
| 1985 | 0906 | 1985 | 1006 | --x- | ---- | ENPS | FLDS | FDMT | 2 |
| 1985 | 1202 | 1985 | 1203 | --x- | ---- | ---- | x--- | ---- | 1 |
| 1985 | 1229 | 1986 | 0208 | --x- | ---- | ---- | x--- | ---- | 2 |
| 1986 | 0318 | 1986 | 0405 | --x- | ---- | ---- | x--- | ---- | 2 |
| 1986 | 0713 | 1986 | 0713 | --x- | ---- | ---- | x--- | ---- | 1 |

| Start Year | | Stop Year | | \multicolumn{6}{c}{Eruption Characteristics} |
|---|---|---|---|---|---|---|---|---|---|

| Start Year | | Stop Year | | CERF | SIGC | ENPS | FLDS | FDMT | VEI |
|---|---|---|---|---|---|---|---|---|---|
| 1986 | 1112 | 1986 | 1113 | --x- | ---- | ---- | x--- | ---- | 1 |
| 1986 | 1125 | 1986 | 1127 | --x- | ---- | ---- | x--- | ---- | 1 |
| 1986 | 1206 | 1987 | 0106 | --x- | ---- | ---- | x--- | --x- | 0 |
| 1987 | 0106 | 1987 | 0210 | --x- | ---- | ---- | x--- | ---- | 2 |
| 1987 | 0610 | 1987 | 0629 | --x- | ---- | ---- | x--- | ---- | 1 |
| 1987 | 0719 | 1987 | 0720 | --x- | ---- | ---- | x--- | ---- | 0 |
| 1987 | 1106 | 1987 | 1108 | --x- | ---- | ---- | x--- | ---- | 1 |
| 1987 | 1130 | 1988 | 0101 | --x- | ---- | ---- | x--- | ---- | 2 |
| 1988 | 0207 | 1988 | 0402 | --x- | ---- | ---- | x--- | ---- | 2 |
| 1988 | 0518 | 1988 | 0801 | --x- | ---- | ---- | x--- | ---- | 2 |
| 1988 | 0831 | 1988 | 1026 | --x- | ---- | ---- | x--- | ---- | 2 |
| 1988 | 1214 | 1988 | 1229 | --x- | ---- | ---- | x--- | ---- | 2 |
| 1990 | 0118 | 1990 | 0119 | --x- | ---- | ---- | x--- | ---- | 2 |
| 1990 | 0418 | 1990 | 0508 | --x- | ---- | ---- | x--- | ---- | 2 |
| 1991 | 0719 | 1991 | 0721 | --x- | ---- | x--- | x--- | ---- | 2 |
| 1992 | 0827 | 1992 | 0923 | --x- | ---- | x--- | x--- | ---- | 2 |

Key:

| | |
|---|---|
| C = Central crater eruption | S = Submarine eruption |
| E = Eccentric (parasitic) crater | I = Island forming eruption |
| R = Radial fissure eruption | G = Subglacial eruption |
| F = Regional fissure eruption | C = Crater lake eruption |

| | |
|---|---|
| E = Explosive (normal) | F = Lava flow(s) |
| N = Nuees ardents, pyroclastic flows | L = Lava lake eruption |
| P = Phreatic explosions | D = Dome extrusion |
| S = Solfataric activity | S = Spine extrusion |

| | |
|---|---|
| F = Fatalities | VEI = Volcanic Explosive Index |
| D = Destruction of land, property | (See Appendix B) |
| M = Mudflows (lahars) | |
| T = Tsunami (giant sea waves) | |

[1] Source: *"Volcanoes of the World"*, L. Simkin & L. Siebert, Washington, D.C.: Smithsonian Institution, 2nd Edition, 1993.

# APPENDIX D

## FORECAST ANALYSIS[1] FOR VOLCANO PITON de la FOURNAISE

Location:   Réunion Island, France Indian Ocean
       21.35S Latitude  055.17E Longitude
       2,631 m  (8,632 ft.)

| | | |
|---|---|---|
| Number of Eruptions | = | 171 |
| First Eruption Year | = | 1640 |
| Last Eruption Year | = | 1992 |
| U (Eruption Rate/Yr) | = | .4857955 |
| Variation | = | 0.000 |
| Pr(0) | = | .6152076 |
| Pr(1) | = | .2988651 |
| Statistical Forecasted Eruption Year | = | 2097 |
| Forecasted Eruption Year @ $\geq 50\%$ | = | 1993 |
| Forecasted Eruption Year @ $\geq 95\%$ | = | 1998 |

[1] As of 1 January 1996

Note: Piton de la Fournaise is classified as a shield type volcano and does not necessarily follow a Poisson distribution in its eruption characteristics. This analysis, performed by "ERUPTION", is based on these assumptions.

978-0-595-41775-9
0-595-41775-2